REHABILITATED BUILDINGS

LINKS

REHABILITATED BUILDINGS

Edition 2010

Author: Roberto Bottura, architect
Graphic design & production: Roberto Bottura, architect
Collaborator: Oriol Vallés, graphic designer
Text: contributed by the architects, edited and translated by Jay Noden

© Carles Broto i Comerma

Jonqueres, 10, 1-5
08003 Barcelona, Spain
Tel.: +34-93-301-21-99
Fax.: +34-93-301-00-21
info@linksbooks.net
www.linksbooks.net

REHABILITATED BUILDINGS

LINKS

007 Introduction

008 Cristian Cirici & Carles Bassó

Vapor Llull

020 Roberto Menghi

Castle in Lodigiano

030 Messana O'Rorke Architects

The Pent Tank House

040 Mónica Pla

Loft Guillermo

050 Albori Associati

House in Appenninos

060 Armand Solà

Loft in Sabadell

068 carlorattiassociati

Cinato Penthouse

076 Recdi8

Loft in Poble Sec

090 Anne Bugugnani + Mónica Pascual

Loft in Lloret

106 Rüdiger Lainer

Penthouse Seilergasse

112 Uras + Dilekci Architects

Misir loft

Michèle & Miquel 120

Apartment in Virreina

MARC (Michele Bonino, Subhash Mukerjee) 134

Apartment in Torino

Brookes Stacey Randall 142

Lowe Apartment

Manel Torres (IN Decoración) 148

Loft in an old factory

Jahn Associates Architects 156

Grant House

MARC (Michele Bonino, Subhash Mukerjee) 168

House in Via Barbaroux

Giovanni Scheibler 178

Loft Conversion in Zurich

Jay Kammen, KAMMEN 186

Pablo Loft

Marco Savorelli 196

Nicola's Home

Otorino Berselli & Cecilia Cassina 206

Restoration in Manerbio

Guido Canali 218

Water Mill

Non Kitch Group bvba 228

Architecture and lifestyle

Introduction

When it comes to renovating, there is no set of "right" or "wrong" criteria. Each project brings with it a unique combination of challenges, problems, strengths and weaknesses which have possibly never been seen before. On historic buildings, how much of the old should be conserved? How far should the renovation either imitate or diverge from the original? What sort of new technologies and materials are compatible with old structures and finishes? These are just some of the questions which inevitably arise when renovating; and the best architects understand that the answers that apply in one project can never be re-used in subsequent programs. Everything must be reevaluated in light of the new challenges posed by new projects.

The results of our search for some of the most exemplary work currently seen in the field of renovating are varied. Defunct factory buildings, centuries-old stone structures and elegant vaulted spaces are but some of the challenges facing the designers in this collection - all resolved with skill and artistry. With this selection, we also hope to present a well-rounded vision of each project. To meet this end, we have endeavored to touch upon every aspect in the planning and renovation processes. After all, technical know-how is just as important as artistic vision in any project.

From conception to completion, we have included information on the materials used and construction processes in order to complement the ideas of the contributing architects. Finally, since nobody is in a better position to comment on these projects than the designers themselves, we have included the architects' own comments, conceptual inspiration and anecdotes.

Therefore, we trust that we are leaving you in good, expert hands and that this selection of some of the finest, most innovative architectural solutions in the world will serve as an endless source of inspiration. Enjoy!

Cristian Cirici
& Carles Bassó

Vapor Llull

Barcelona, Spain

Photographs: *Rafael Vargas*

The Vapor Llull (a steam-driven factory), in an old industrial district of Barcelona, consisted of a set of buildings dating from the early 20th century which had been devoted to manufacturing chemical products.

The basic structure of the complex consisted of a long ground floor plus two floors, the highest of which had a sloping roof supported by a structure of wooden trusses. The complex also included a series of auxiliary premises adjoining the main building and a magnificent brick chimney over 30 m (100 ft) high that was part of the steam engine that powered the factory. The architects decided to conserve the chimney in order to keep alive a symbol of a time in which the whole district was full of steam-driven factories. The most suitable property for conversion into loft dwellings was the long main building. In order to create an open, private space and provide a one-vehicle parking space for each of the eighteen units into which the scheme was subdivided, a series of auxiliary buildings were demolished. To give independent access to each module of approximately 90 sqm (970 sqft), three sets of vertical communication elements were introduced, each with a stairwell and a panoramic elevator. Their formal expression gives the appearance of silos covered with enameled corrugated steel. On the outside the main building was painted with silicate paint applied directly to the bricks, which were first stripped of their render.

In the interior, it was decided to leave the spaces free and unfinished; so that each loft could be arranged according to the wishes of the different designers of interiors that were chosen to finish off the scheme. The layout and decor of this loft are by Inés Rodríguez. It is a two-level apartment in which a mezzanine houses the bedroom and a bathtub. It is a curious habitat in which the light and the space create an atmosphere of elegance.

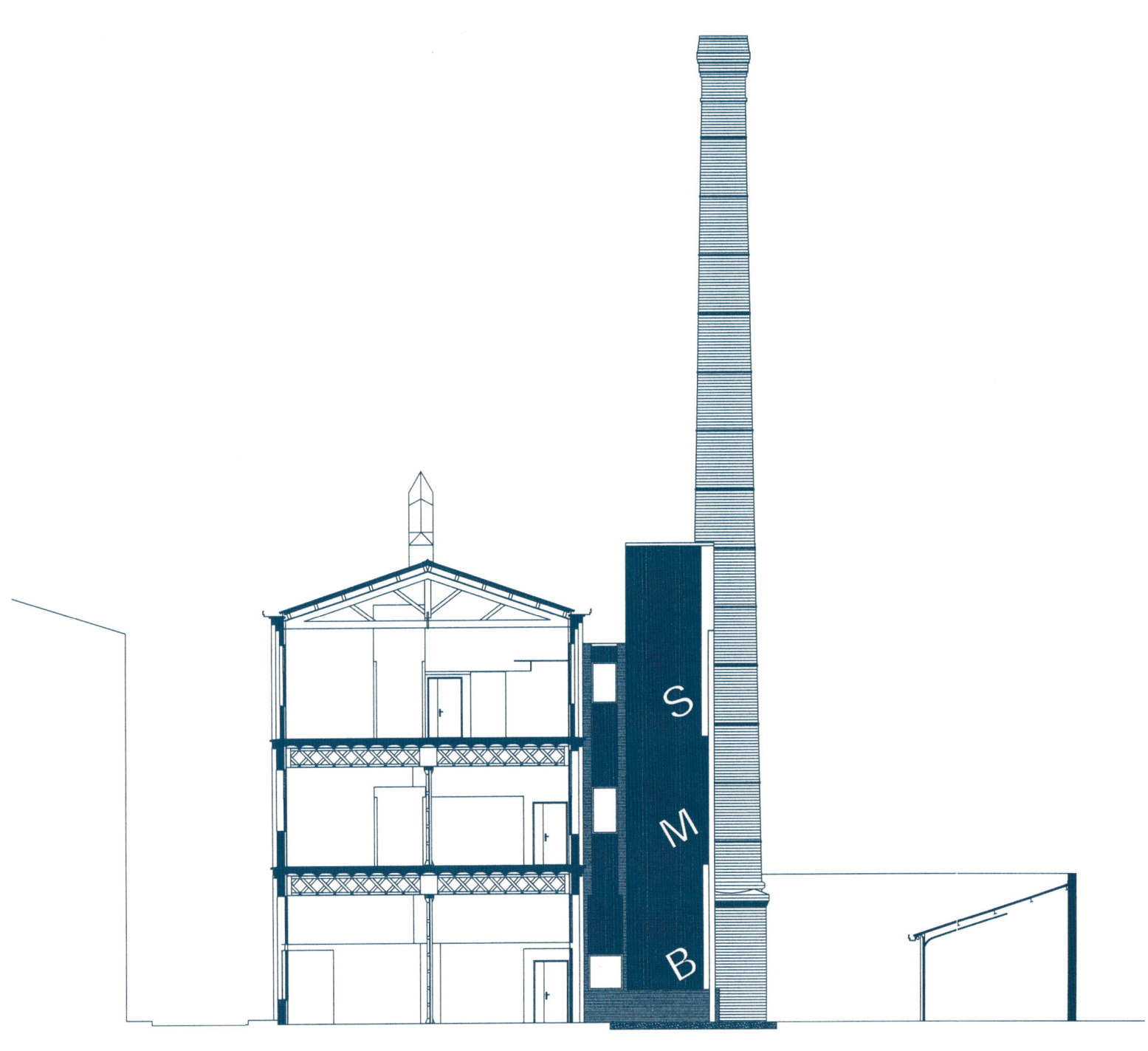

1-2
3-4
S
5-6

S
M
B

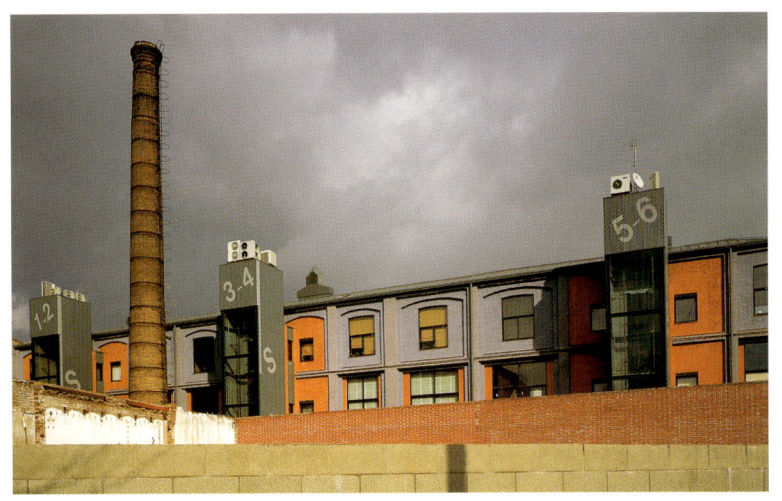

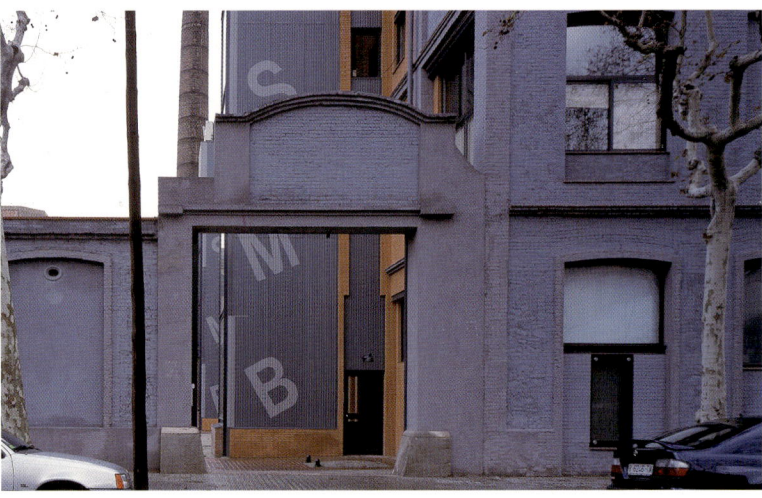

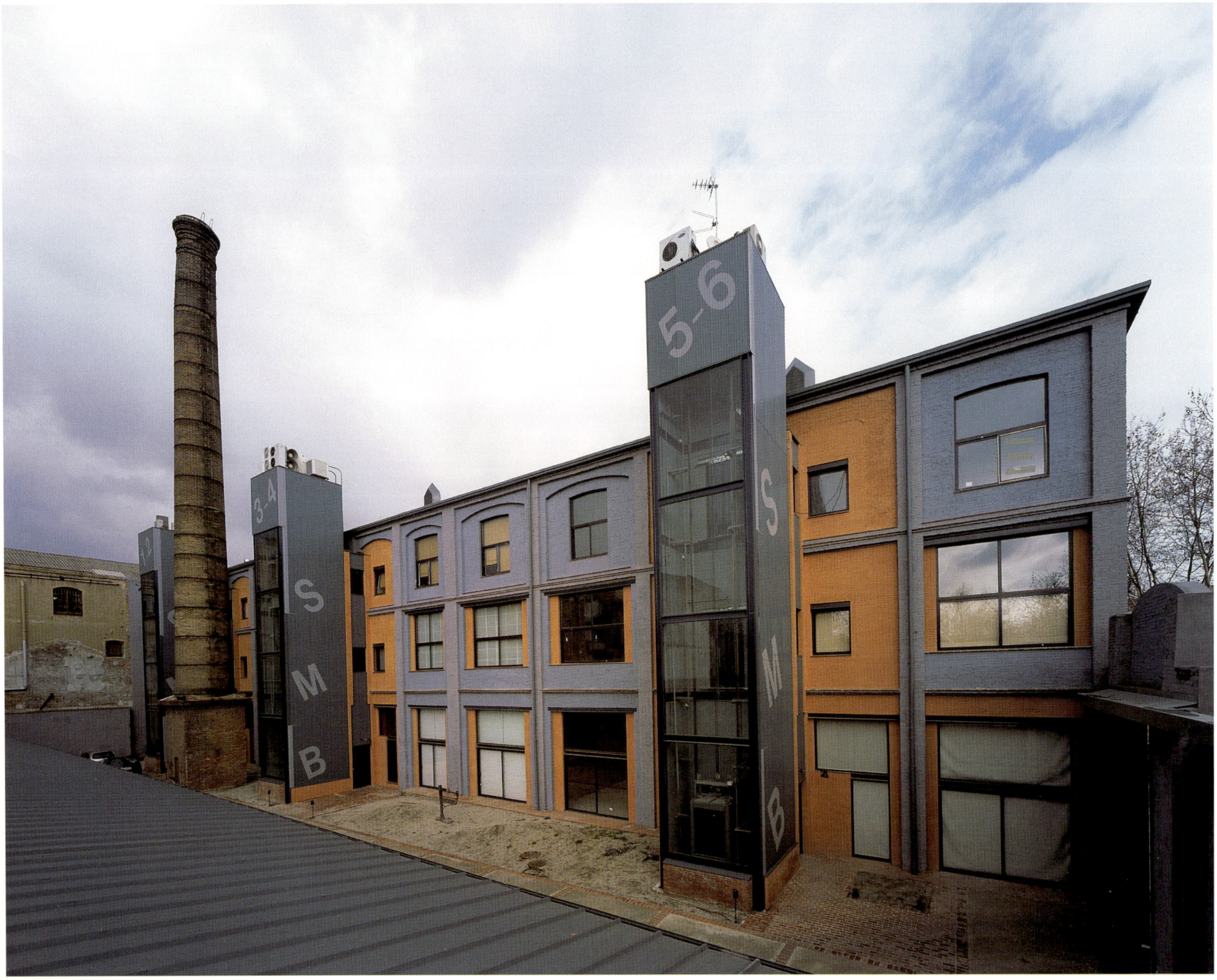
1-2
3-4
S
M
B
5-6
S
M
B

In this apartment large spaces with bright, clear walls are predominant. The most outstanding features at first sight are the wooden trusses in the whole dwelling and the polished concrete floor.

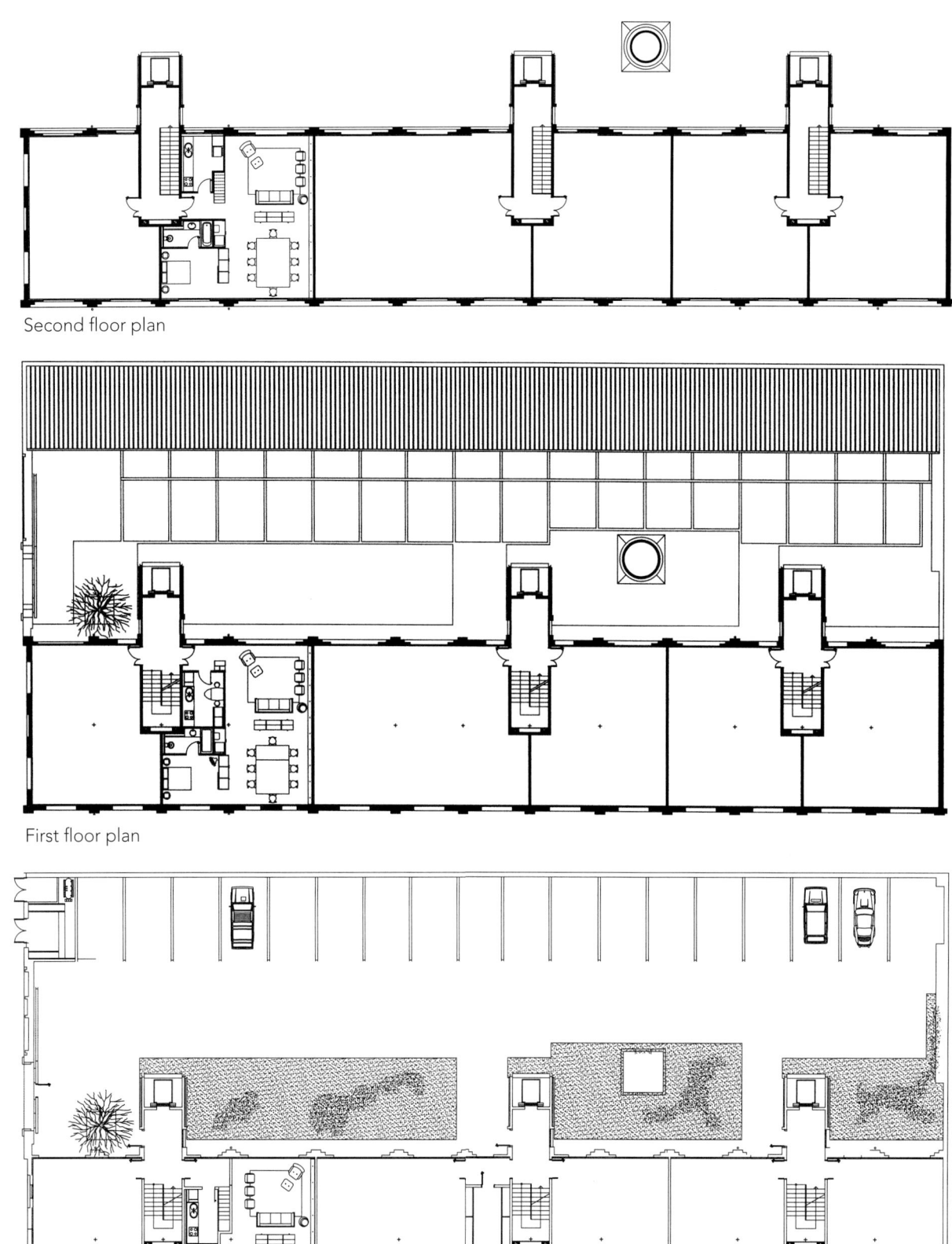

Second floor plan

First floor plan

Ground floor plan

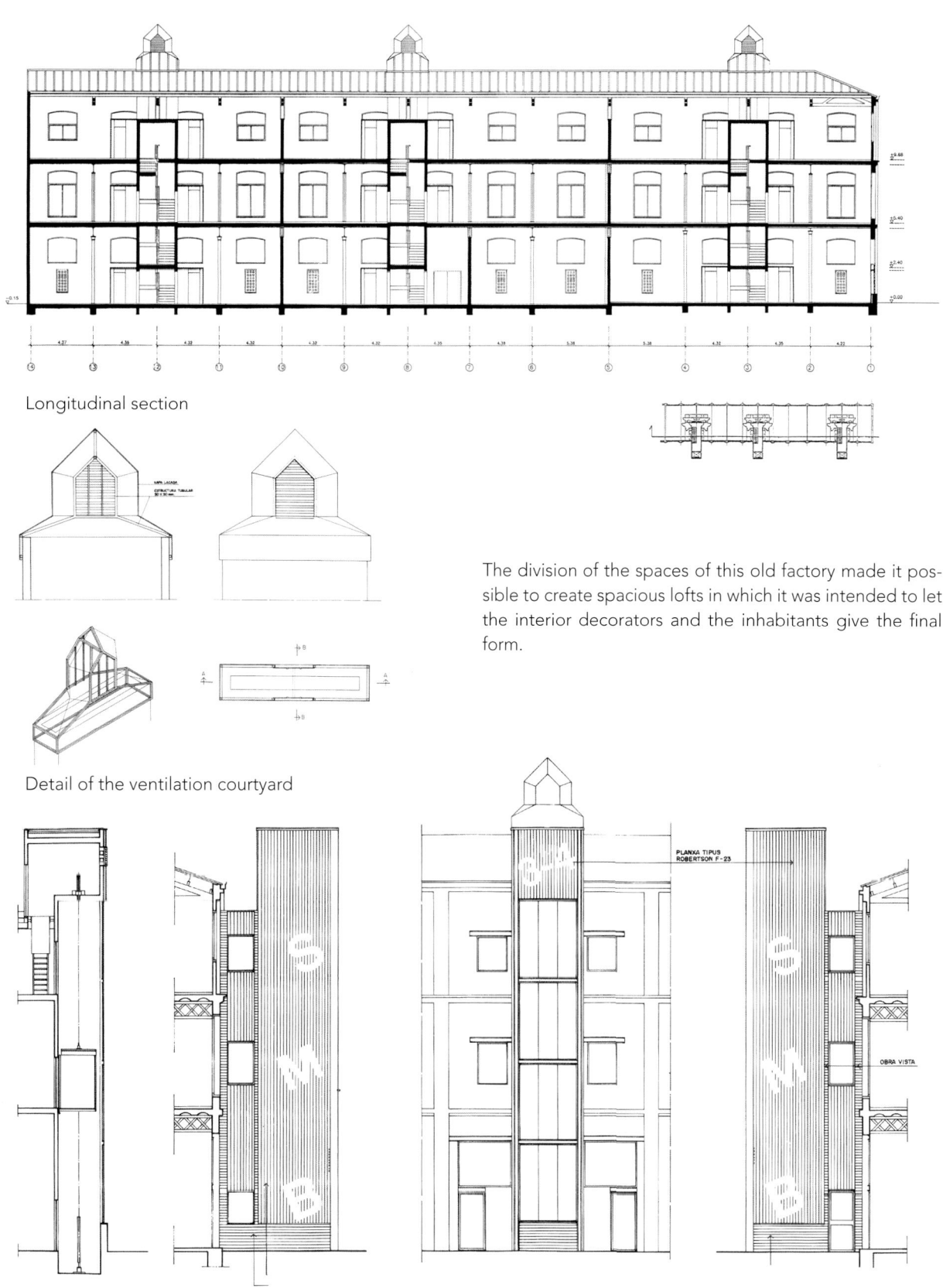

Longitudinal section

Detail of the ventilation courtyard

The division of the spaces of this old factory made it possible to create spacious lofts in which it was intended to let the interior decorators and the inhabitants give the final form.

Panoramic detail of elevator windows

In order to create an open, private space and provide a one-vehicle parking space for each of the eighteen units into which the scheme was subdivided, a series of auxiliary buildings were demolished.

Roberto Menghi

Castle in Lodigiano

Lodigiano, Italy   Photographs: *Melina Mulas*

The aim of the project was to restore the north-western part of the Castle, modernizing the interior and making it fully habitable without spoiling the unique character it had taken on over the years.

New connections have been made between the ground and the first floors through the insertion of two new steel spiral staircases with steps in solid bay oak; the existing straight staircase in stone was resurfaced. The horizontal structures on the first floor, original beams and shelves in bay oak, were restored and reinforced with a special procedure based on resin and metal inserts. The existing terracotta tile floors were levelled off and integrated with new hand-made tiles of the same size and in the same clay (which is still found in the area) as the old ones. To solve the problem of rising damp, which has been penetrating the wall for centuries, an insultation system using "active electro-osmosis" has been adopted with excellent results.

The project also creates an intermediate level between the ground and the first floor, in part adapted as a study-library facing over the main hall, in part used as a service area: cloakroom, laundry, etc. The 6 m (20 ft) ceiling height typical of the age has been maintained for the entrance hall, part of the main hall and the whole kitchen. The intermediate level has been made, like the stairs, using steel structures, but with flooring in Swedish pine.

The two stairways and this last space were designed and constructed with materials that contrast with the original context, in order to highlight their super-structural and "removable" nature.

Bedrooms, bathrooms and closets have been fitted into the space situated above the pointed arches. The outer wall of the bedrooms overlooking the courtyard has been moved back from the rest of the façade in order to make room for a long (approximately 2 m, or 6.5 ft, in width) balcony, above the arches.

The Castle looks like a fortress with a moat, having surrounding walls and a shortage of apertures, with a consequent shortage of light and air in the rooms. This problem was solved by reopening some of the original apertures that had been walled over and making some new horizontal slits under the western eaves.

The roof has been restored and insulated, maintaining the existing cover in bent hand-made tiles.

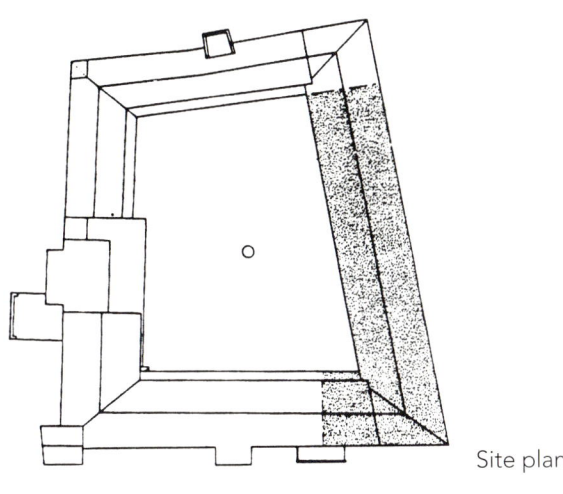

Site plan

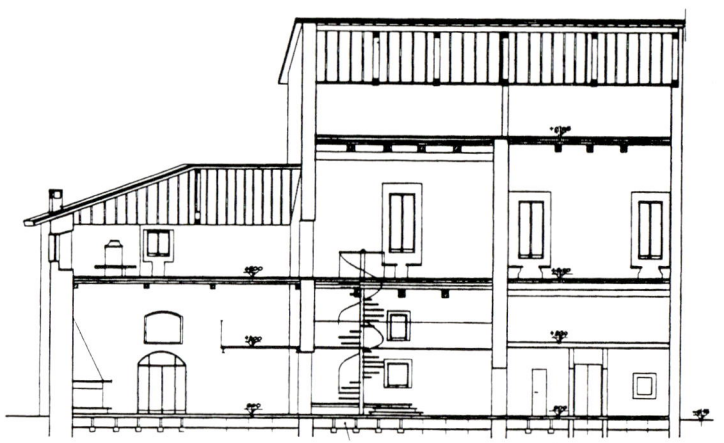

Section A-A

View of the large courtyard of the castle, with large porticoes providing spaces protected from direct sunlight.

Part of the work consisted of inserting a new level between the ground floor (6 m, or 20 ft, high) and the upper level, and establishing a new vertical connection by means of a new spiral staircase with solid oak steps.

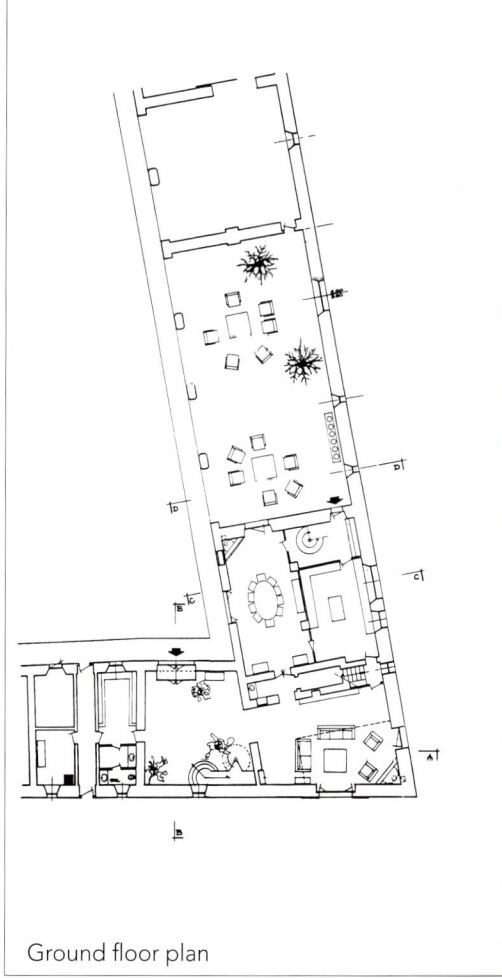

Ground floor plan

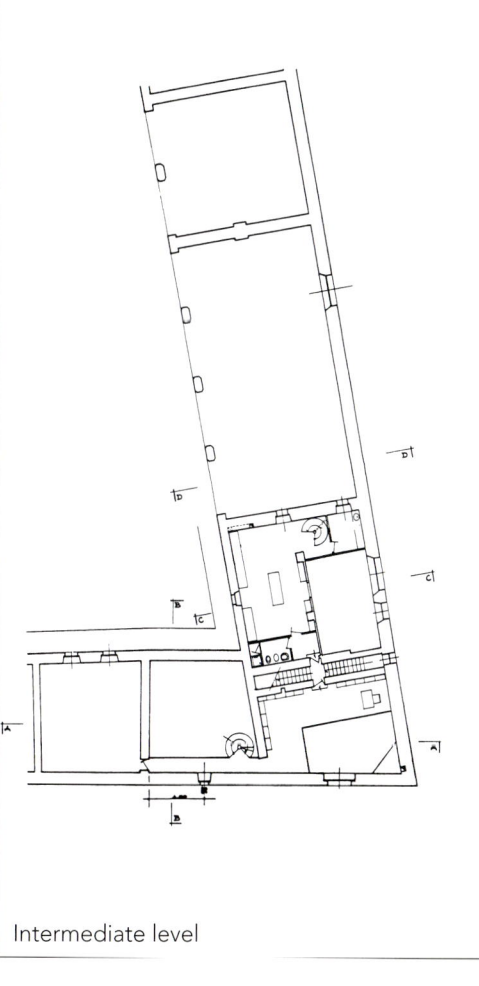

Intermediate level

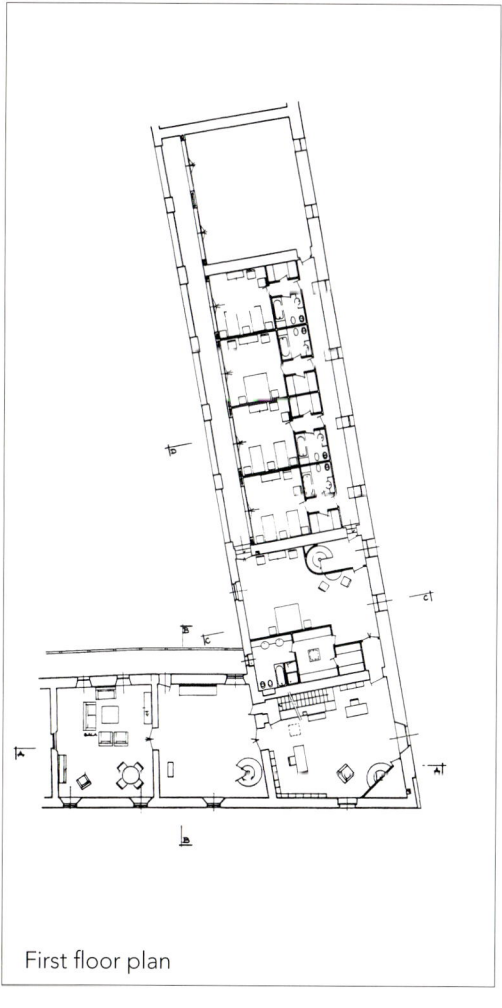

First floor plan

The original loadbearing structure made of wooden trusses has been treated with resins for conservation and reinforced with metal elements.

Messana O'Rorke Architects

The Pent Tank House

New York, USA   *Photographs: Elizabeth Felicella*

The project involved the renovation of an existing loft apartment and the development of an existing sprinkler tank house into an urban retreat.

The apartment's simple layout was maintained, but reconfigured so that the bedrooms had individual access to a shared bathroom rather than it being accessed from the living space, (a powder room in the entry way meant that guests would not need to access the bathroom). The sky lit bathroom doubled in size and was developed to have a serene spa like quality, with a continuous stone floor, polished waterproof plaster walls, and clear glass shower enclosure. In the living room and bedrooms new storage was introduced at every possible point behind hidden flush lacquer panels. The kitchen was resurfaced and the maple wood floors were lightened and refinished. A custom stainless steel spiral stair replaced an ugly painted steel stair with wood treads. The client anticipated that a new stair would encourage him to ascend more frequently to his roof deck and Tank House.

A tree house perched high in a city of towers and skyscrapers. The tank house was conceived as the quintessential retreat, a room for reading, relaxing and listening to music.

The refurbished roof deck was given new trees and landscape; it had existed for a number of years, over-shadowed by a looming tar covered rotunda occupied by an enormous cast iron sprinkler tank. The removal of this tank and the introduction of a new structural frame to support the crumbling terracotta walls of the rotunda were essential difficulties of this project. The building was shored-up by an external wood frame and the tank was slowly cut into manageable pieces with blowtorches. Once removed the true proportions of the internal space were revealed and it was tempting to leave it as a raw industrial space, but the program requirements and in particular the need for the room to be usable year round meant moving forward as planned.

A 3.5 m (12 ft) tall window was cut into the east side of the space and the new window looked out onto the roof deck plantings.

A circular skylight was introduced into the center, casting an ethereal light into the space. The floor of maple matched the apartment and was segmented into removable panels providing access to storage space below.

The sky lit bathroom doubled in size and was developed to have a serene spa like quality, with a continuous stone floor, polished waterproof plaster walls, and clear glass shower enclosure. In the living room and bedrooms new storage was introduced at every possible point behind hidden flush lacquer panels.

Mónica Pla

Guillermo's Loft

Barcelona, Spain   Photographs: José Luis Hausmann

Before it was renovated, this old apartment located in Barcelona's Ciutat Vella quarter, featured a lot of rooms and little natural light. The deteriorated and neglected space has been transformed with the goal of creating a luminous and spacious residence in just one environment.

The first step was to tear down various walls and to open up -structure permitting- a number of windows to permit the entrance of natural light. With no walls left to divide the space, the designer managed to provide the bedroom with intimacy and to establish a difference between the kitchen and the living area. To separate the bedroom from the main room and still maintain spatial continuity, designer Monica Pla placed the bathroom between the two. Accessible from the bedroom, the bathroom contains a shower which is contained in a box with partial walls. The sink, mounted on one of the exterior faces of the box, is completely integrated into the bedroom space. The toilet is situated next to the entrance door and a sliding door completely isolates it from the central living area.

To obtain harmony in the apartment, the furnishings were carefully selected. On one of the lateral walls, a piece of IKEA furniture unites the entire space, from the kitchen-dining room to the living room. As a result, the kitchen is integrated with the living room, creating a warmer environment. In the living room, Pla exposed a brick wall and managed to define the room's independence by means of a sofa upholstered in brown canvas from DOM. The dining room table is from Pilma. The lighting, also from Pilma, contributes to the warmth of the space, as does the wooden floor.

By studying the placement of the pieces of furniture and other objects that make up the home, the interior designer achieved a functional and modern renovation that floods a remarkable, diaphanous space with light.

Light colors, used for the furnishings, the hardwood floors, and the kitchen cabinets, present a uniform and spacious atmosphere towards the hallway.

To incorporate the shower into the center of the space required the use of a piece of furniture to divide the apartment's two main areas. A small stair leads up to the shower and the new installations are hidden underneath this platform, which reinforces our perception of it as a foreign element placed on top of the original floor.

A shower has been placed in a newly-created volume that separates the bedroom from the living room, while maintaining visual continuity. This volume is slightly raised, with the necessary technical installations set beneath.

Albori Associati

House in Appenninos

Montese, Italy

Photographs: *Matteo Piazza / Albori*

This old farmhouse of an uncertain age, which has been remodelled several times, is set on a hillside of the Apennines in the region of Módena.

The idea of its new owners was to reduce the bathrooms and the night area as far as possible in order to leave the rest of the spaces open and at the disposal of the house. Because it was a building without partitions, in which the interior space coincided with the structure, most of the rooms were on different levels. This pure relationship between space and structures was so beautiful that it was decided not to alter it.

Therefore, none of the rooms were divided and the new volumes are characterized by the type of material of the furniture inside them: wood, bricks and stone help to define the uses of the different rooms. The main barn had a certain majesty, with an irregular geometry, a floor area of 48 sqm (515 sqft) and a roof that is 6 m (20 ft) high at its highest point and 2.4 m (7.8 ft) at its lowest point. After the demolition of the floor located between the ground floor and the first floor, a large room of double height perfectly illuminated by a long thin skylight was created. The darkest area in the dwelling was illuminated thanks to the courtyard located at the end of the old barn. In order to adapt this impressive building, originally designed for agricultural use, it was necessary to redesign its lighting and ventilation, without forgetting the enormous possibilities of making openings due to the excellent situation.

The darkest area of the dwelling was illuminated thanks to the courtyard located at the end of the old barn. The beautiful views of the valley and the hills surrounding this house were accentuated by the creation of a large terrace in the east wing of the dwelling and by the creation of new windows. These openings led the architects to create a tortuous and circular layout full of transparencies and unexpected views through the rooms and toward the exterior.

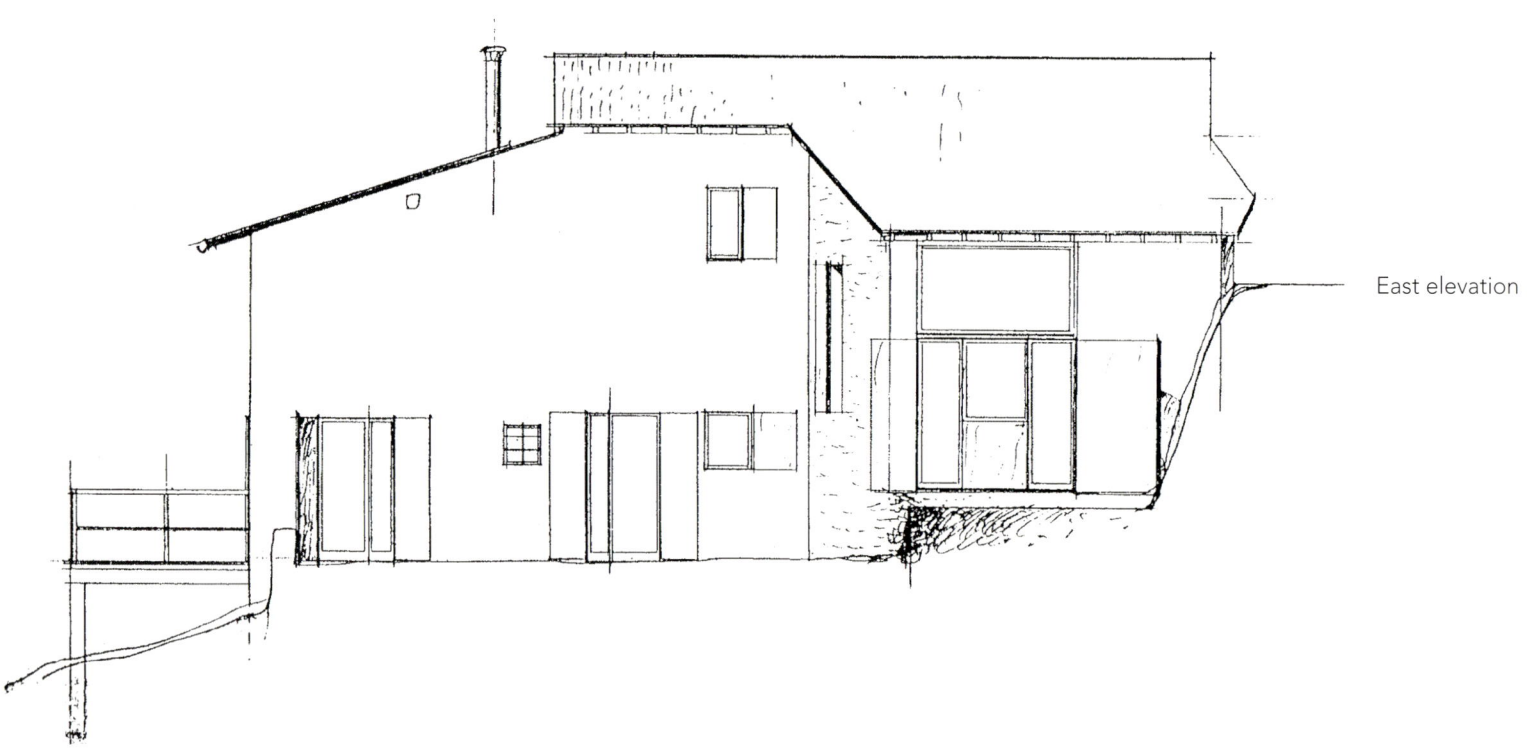

East elevation

On the exterior of the dwelling, with one of its main walls sculpted from bare stone, wood and iron were chosen to follow the style of the agricultural buildings in the area.

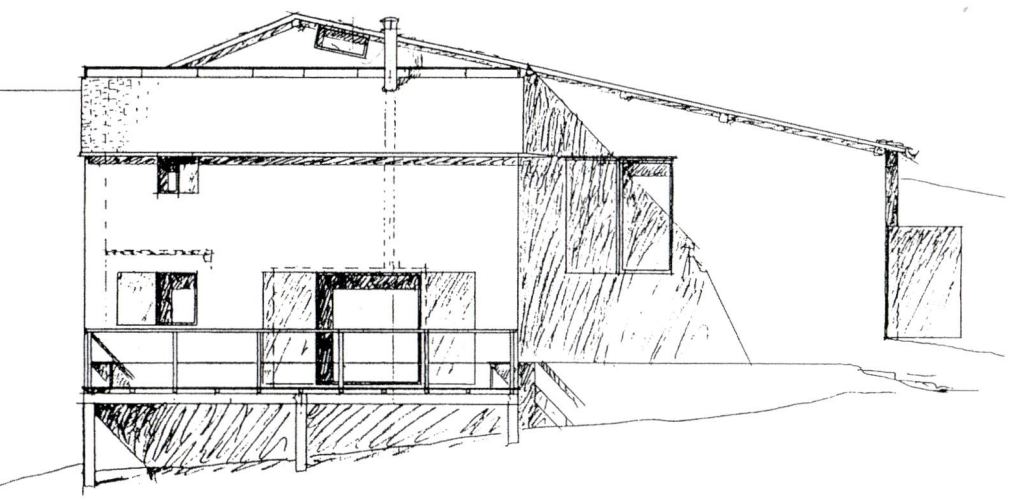

South elevation

In the large living room, the larch wood combines perfectly with the fireplace built with stones taken from the demolition rubble. A staircases rises above the fireplace, with a small toilet underneath it.

Armand Solà

Loft in Sabadell

Sabadell, Barcelona, Spain

Photographs: Eugeni Pons

Furniture:
Sofa: Perobell
Chairs: Andreu Word
Tables: from the author
Curtains: Cortinova

The project involved adapting an industrial textile building for use as a family home. The building consisted of two floors of completely uninterrupted space, with considerable heights and large windows facing outside.

With these initial characteristics, and given that the lower floor was to maintain its original function, the construction of a mezzanine became the way to use the space under the gabled roof. The dwelling would be organized like a loft, flexibly adaptable to the future requirements of the young couple who were going to inhabit it.

Thus, a twofold structural challenge arose: in the first place, the lower chords of the steel roof trusses prevented the upper part of the space from being used; this was solved by transforming the trusses into rigid frames, with angle irons fixed onto each side, so the chord could be cut out without affecting the structure's stability. These angle irons were made unusually heavy, to deal with the second part of the challenge: the mezzanine floor was to be suspended from the roof structure to avoid building pillars through the floors below. Some connections to the previously existing structure were required to avoid any tendency to swing. The first truss in the series was kept as a nostalgic expression of the buildings original state.

The lower floor only contains the entrance hall, with doors through to the workshop on one side or the storage and parking area on the other. The stairs occupy the same place, freed of constricting blind walls.

The main floor is a wide double-height space with a minimal sanitary space to protect the intimacy of the rest of the floor visually, while the brightness of the colors and the numerous elements guessed at are an enticement to proceed.

The distribution of the various zones follows a longitudinal compositional axis that ends at the two large windows overlooking a square. As from here, the only colors are the natural tones of wood, rust, stainless steel and white, there is a constant insistence on visual connections to the upper floor, either through the double height of the living room, or laterally along the longitudinal axis. This connection is reinforced by the windows, the dimensions of which make them serve both floors. The same goes for the radiators, great vertical panels that reinforce a unitary impression.

At the opposite end to the square, the compositional axis terminates at a discreet access to the terrace of the neighboring building, which belongs to the same family. The mezzanine is a wide area for study and work that, in time, can be partitioned into independent bedrooms when needed.

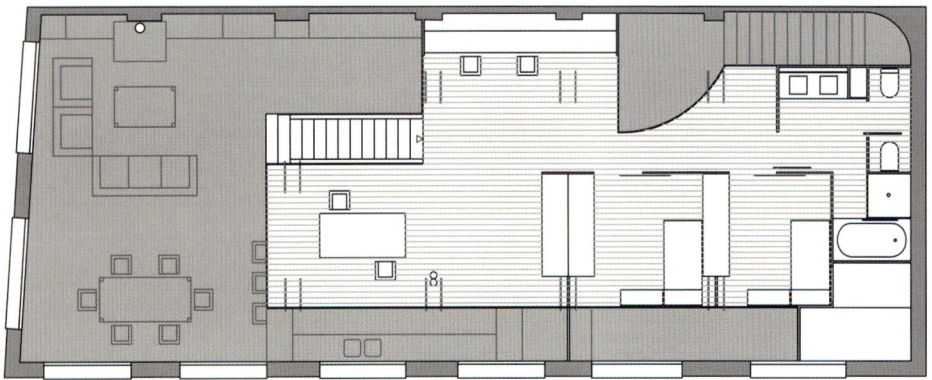

First floor plan

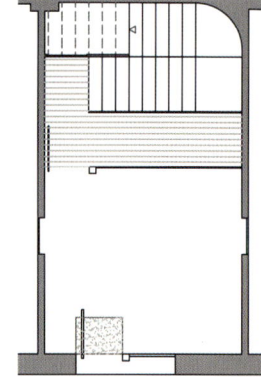

Ground floor plan

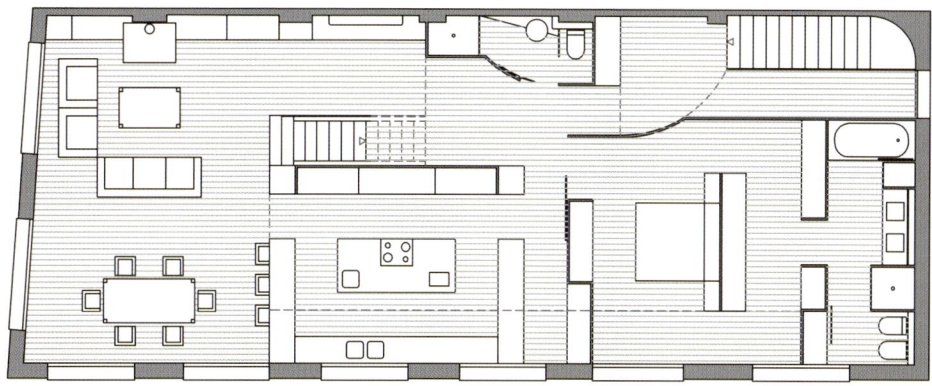

Attic plan

The sobriety of the shapes and the minimalist feeling of its components, in which different materials come together with no transition, together with the austerity of the colors contribute to make this an undeniably beautiful composition. Any domestic item included will establish a contrast, so that in the near future, when the apartment is inhabited, a small intellectual effort will always lead back to the beautiful, serene vision offered by the space in its raw condition.

Natural light is guaranteed everywhere, and the lighting has been designed to create different atmospheres: fluorescent tubes inserted into the roof trusses cast a general diffuse light, while spotlights illuminate specific areas of work, study, cooking etc.

carlorattiassociati

Cinato Penthouse

Turin, Italy

Photographs: *Pino dell'Aquila*

The refurbishment of a penthouse in Turin, Italy, was the opportunity to test innovative Computer Aided Design/ Computer Aided Manufacturing (CAD/CAM) procedures. Digital-minimal: This project sets out to demonstrate how to take advantage of new creative techniques in architecture without producing blobs or funny shapes. Several hundred pieces, each one different from the other, were designed to fit into the existing space. The items where then manufactured out of folded and laser-cut corten steel plate. Finally and magically, all the parts were assembled into a coherent whole within the old walls.

The project was prompted by a new Italian law, which allows owners to convert spaces under the roof into penthouses. A young and liberal Turin couple, both of whom are employed in the high-tech sector, owned the top floor of a multi-story condominium in the city center. They decided to substantially increase the available footage of their property, if necessary by unconventional means. Their initial brief for the reform operation included two new bathrooms and a new staircase to access the roof. The challenging design puzzle that arose was the following: would it be possible to combine all of them into a single architectural element? The bathroom might become a staircase. But could a staircase become a bathroom?

At the entrance level, a large living room opened up, to host the owner's social activities. With a uniform corten steel floor, protected by an epoxy layer, a flexible living room for partying was created. On the upper level, industrial wood flooring was provided, to produce a more private, but equally multifunctional space.

On those occasions in which the new shape met with the original building's structural elements, these were allowed to speak for themselves, expressing a fertile dialogue between the old handcrafted materials and the digital logic of the new forms. The existing structure's functional validity was multiplied by sidestepping the routine of established types of distribution.

Architectural design team:
Walter Nicolino, Chiara Morandini,
Anna Frisa, Carlo Ratti
Structural design:
Studio Vittorio Neirotti, Turin, Italy
Construction:
Impresa Carriere

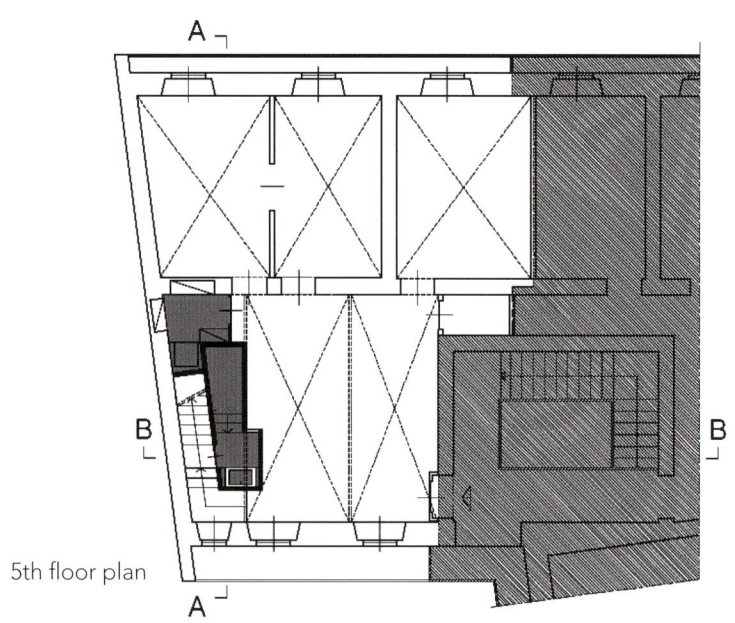

5th floor plan

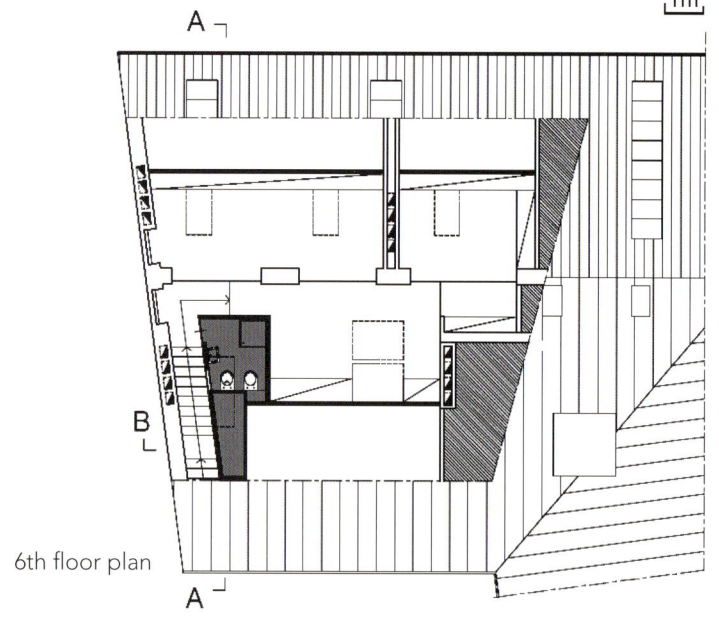

6th floor plan

1m

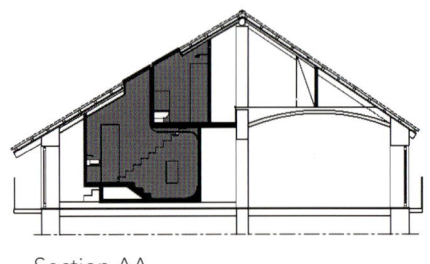

Section AA

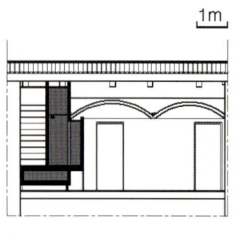

1m

Section AA

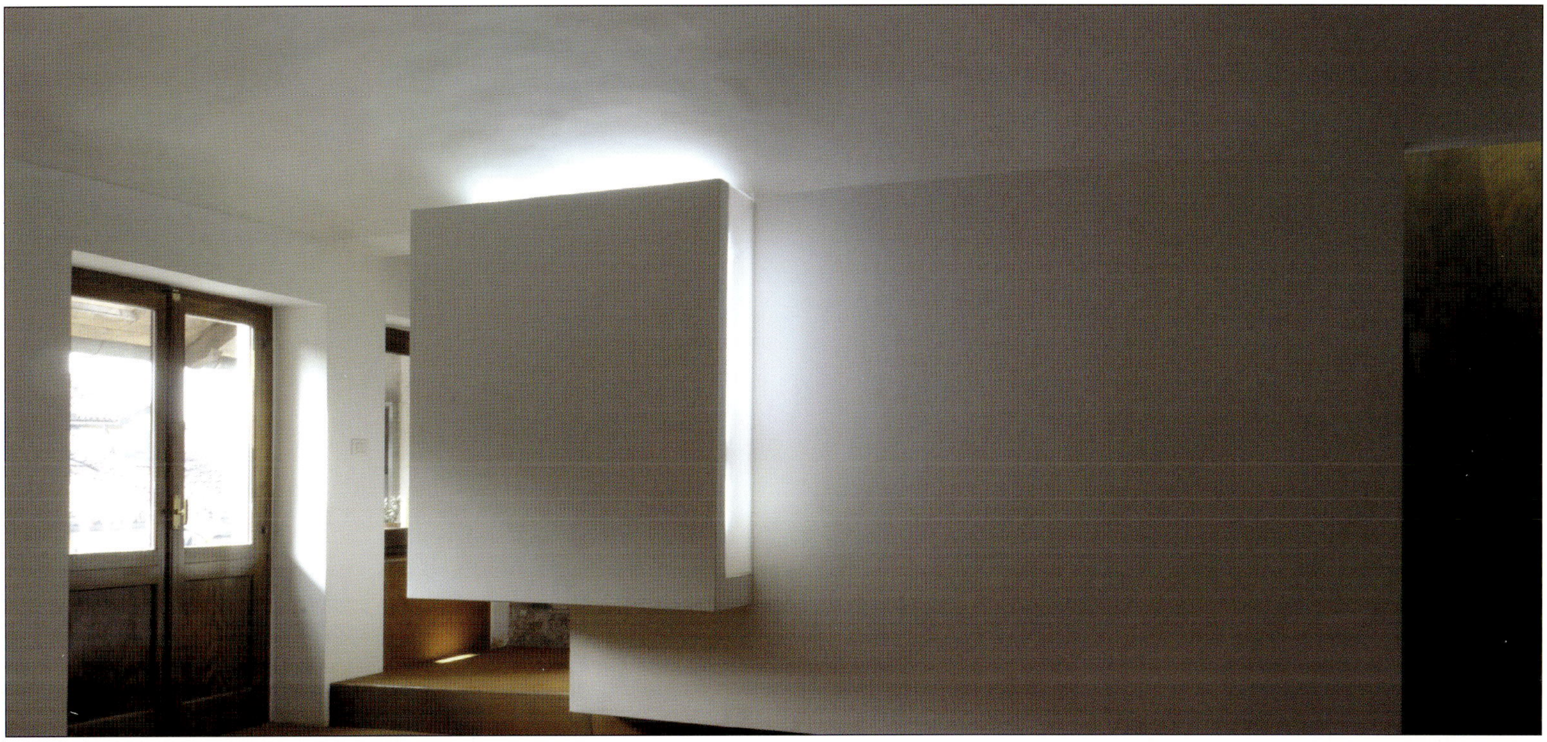

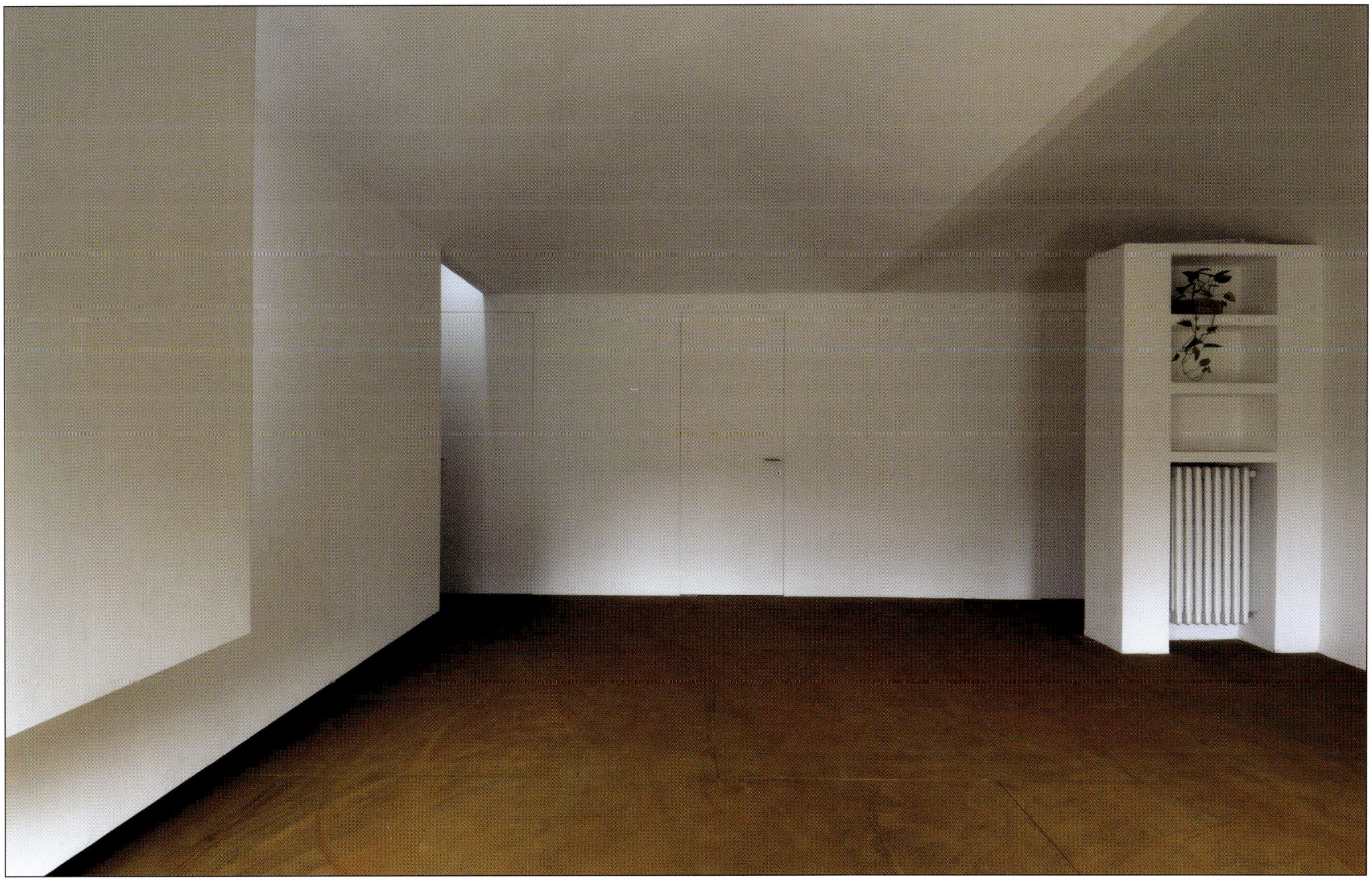

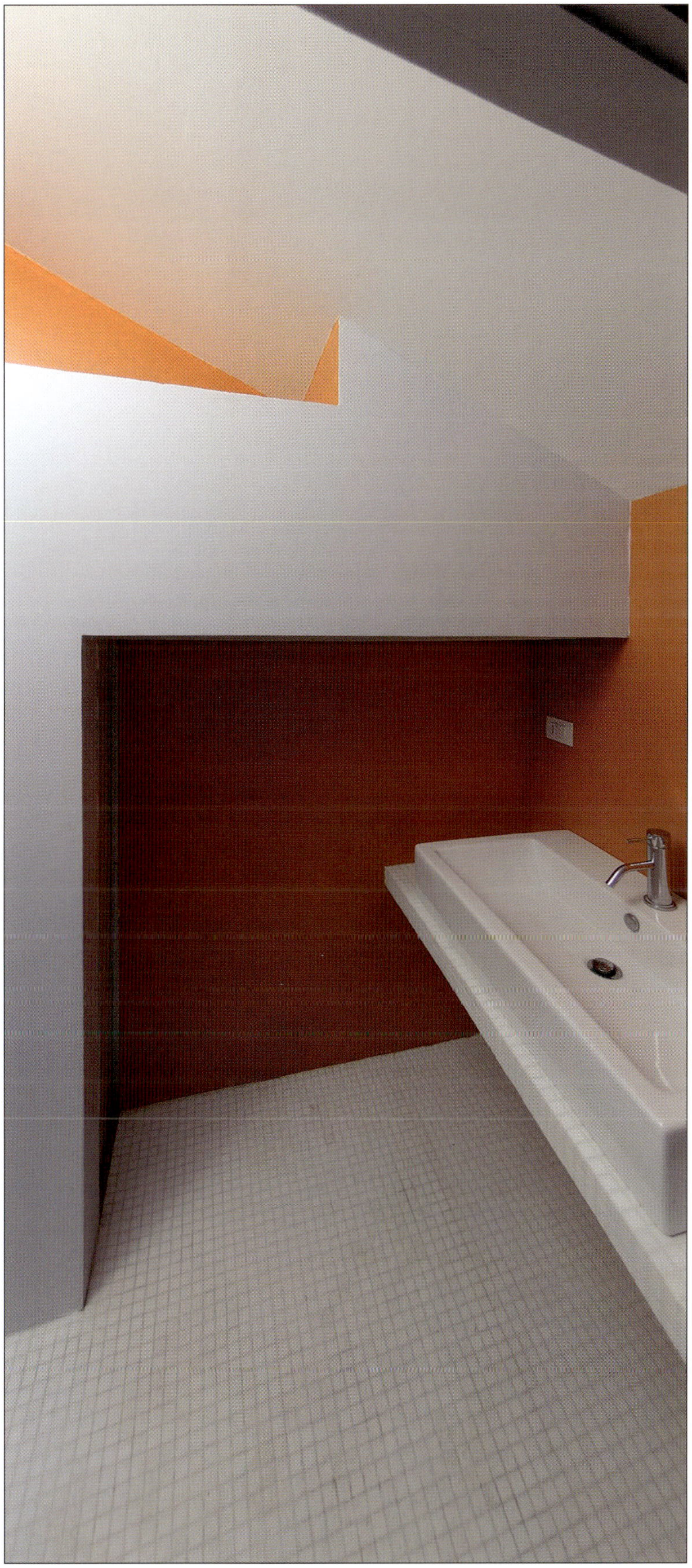

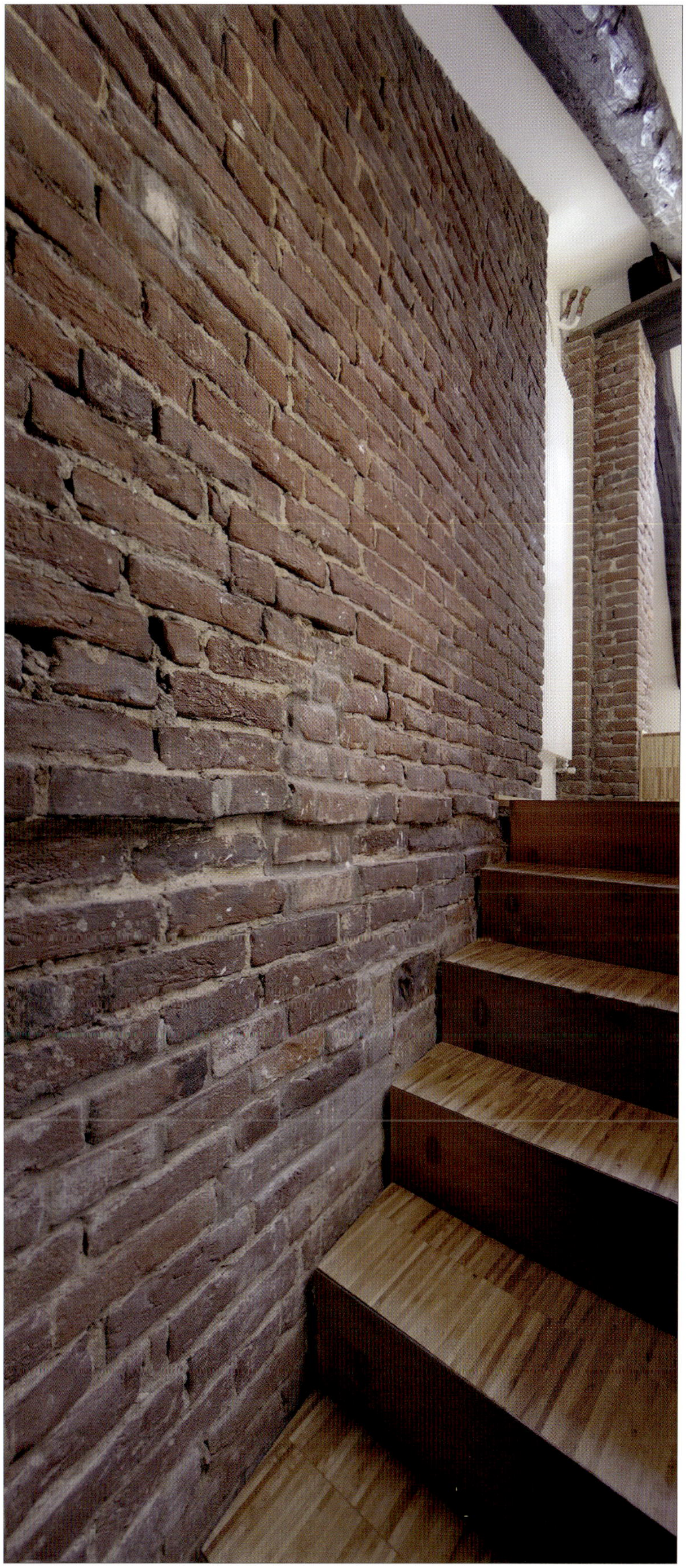

Recdi8

Loft in Poble Sec

Barcelona, Spain — Photographs: *Lane de Castro*

The project called for the conversion of a defunct carpentry workshop, located on a ground floor measuring 100 sqm (1076 sqft) with a ceiling height of 4.5 sqm (15 ft), into a comfortable habitable space.

The floor plan is divided into two main parts joined by a completely white Zen-inspired space. The entire length of the apartment is marked by an interplay of double-height spaces, changes in floor level and lofts.

The entryway is located in the building's communal stairwell; this entrance zone organizes the space so that it communicates directly with a small closet, toilet and the dining room. The former entrance to the workshop is also located here and the old wooden doors have been conserved, protected by a double door of steel and acid-etched glass that conceals views from the street.

It is here that the 4.5 m (15 ft) ceiling height is divided into two spaces via a loft that supports an office and guestroom concealed by teak wood screens. The kitchen is located beneath this loft and the floor had to be lowered in order to achieve the required height. The thick existing stone wall is the unifying element between these two zones. Here, the lighting is provided by fixtures set into the floor.

The flooring is in resin in almost the entire apartment, except for the slate paving in the master bedroom and the oil-treated pine parquet of the guestroom.

On the wall opposite the kitchen is a combination bench/storage unit with a retro look created by the interplay of the wood's color, a reflective gold enameled tiling and lighting fixtures affixed to the wall.

The kitchen gives way to a space finished entirely in white, where soft light is filtered through a skylight in the central zone that was once used as the building's ventilation shaft but which now features a steel and acid-etched glass ceiling and small points of light lying almost flush with the ground. Here also, two stylized windows cast natural light into the main bathroom.

Past this zone, the apartment again opens up into a space with a ceiling at the same height as the first. This space has been split into three areas. First is the TV room, with two '70s-style armchairs and an entertainment unit containing the TV, stereo, CDs, DVD and a bar. A short flight of stairs leads to the first shift in level, where there is a living room overlooking the courtyard on the other side. Here, the openings of the original space have been conserved, and ample light floods this zone. Finally, the master bedroom is set alongside the living room on a platform paved in slate and partially lit by natural light filtering through teak wood screens. The same untreated pine that is used in the other storage units in the house has also been used for the master bedroom closet. A restored antique marble sink now graces the main bathroom, which has been painted in gray stucco for a colder, more industrial feel.

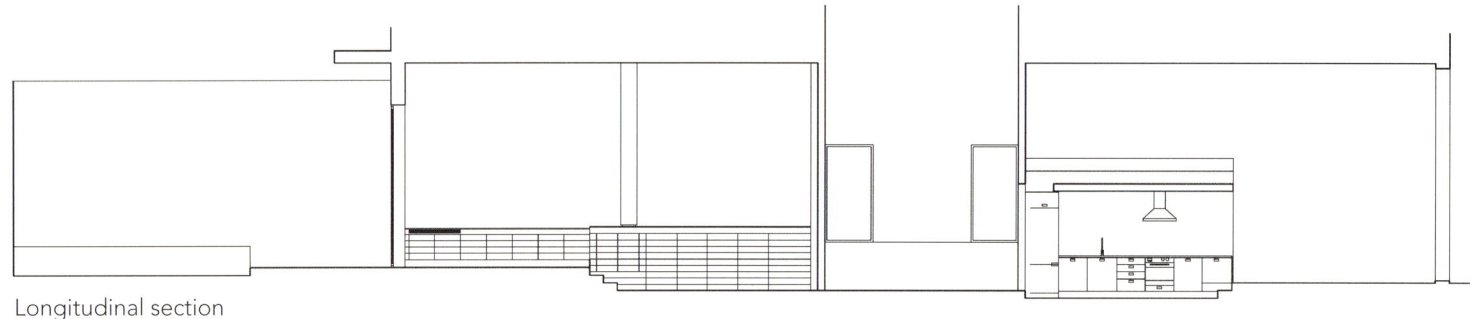

Longitudinal section

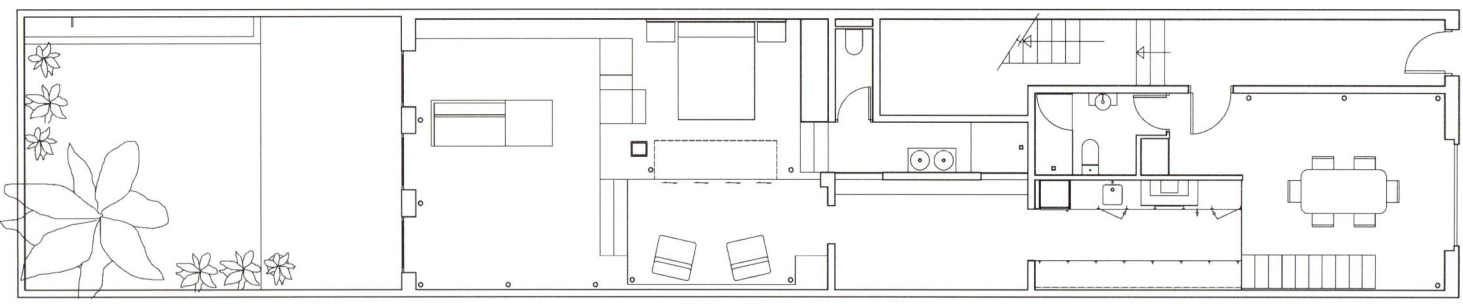

Floor plan

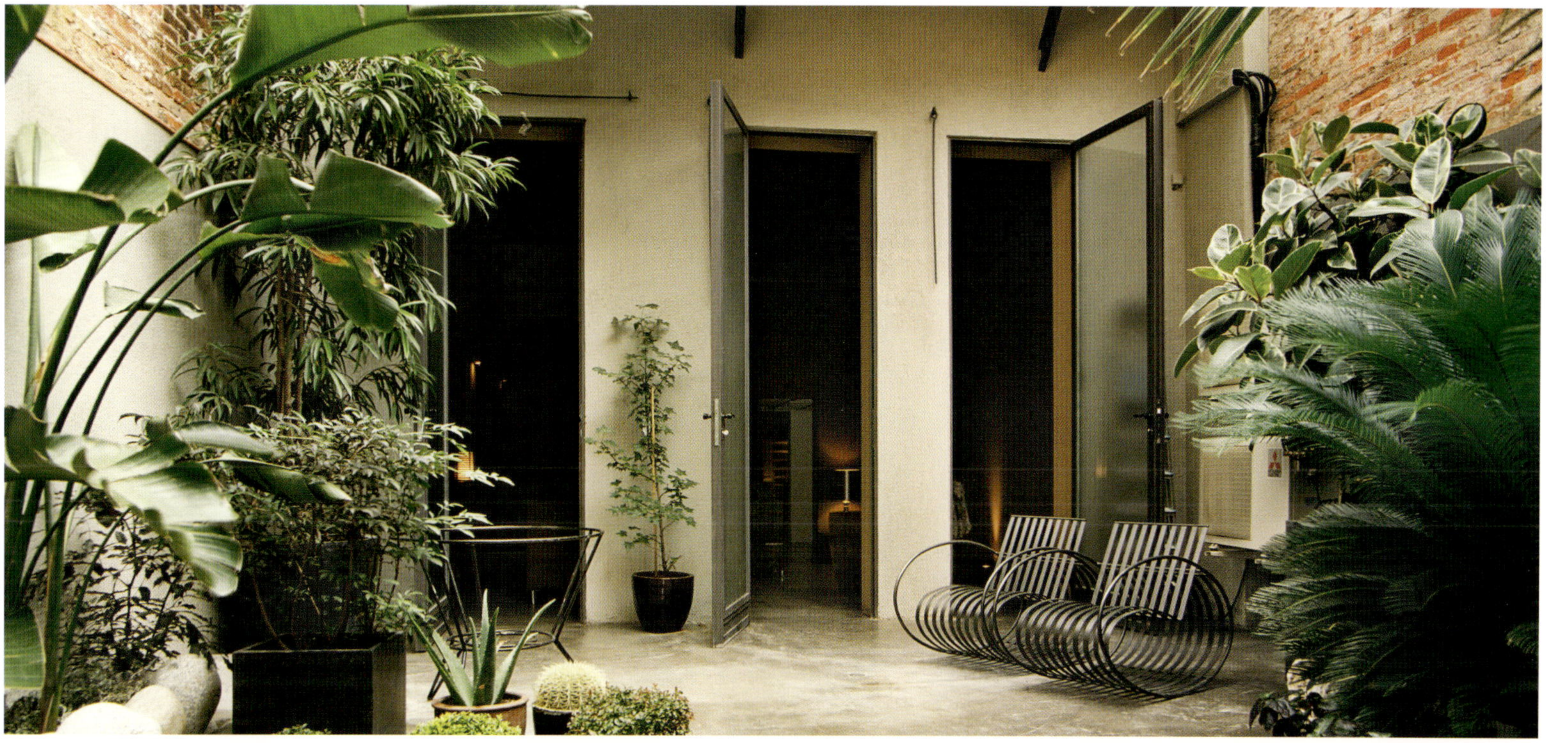

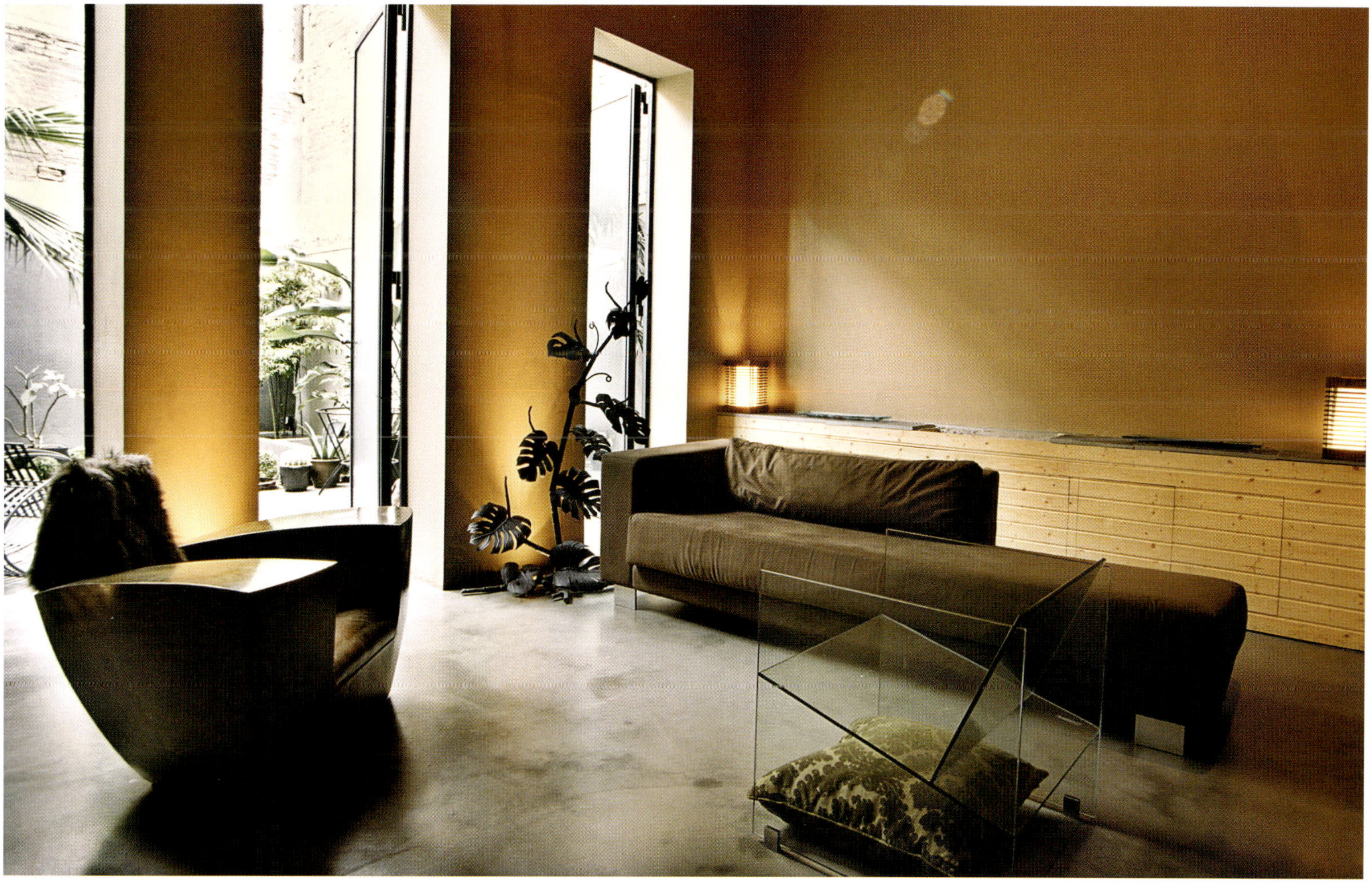

The master bedroom is set alongside the living room on a platform paved in slate and partially lit by natural light filtering through teak wood screens.

A restored antique marble sink now graces the main bathroom, which has been painted in gray stucco for a colder, more industrial feel.

A restored antique marble sink now graces the main bathroom, which has been painted in gray stucco for a colder, more industrial feel.

Anne Bugugnani + Mónica Pascual

Loft in Lloret

Lloret de Mar, Girona, Spain   Photographs: *Eugeni Pons*

This dwelling is in a semidetached stone house, which was once used as a winery and has been abandoned for the past 150 years.

One of the principal architectural interventions for this project consisted of using the back courtyard area to build a vertical service core, which is accessible from each floor. In this way, the new addition was used to add kitchen appliances (refrigerator, washing machine etc) to one level, under a practical glass roof that allowed natural light to penetrate into the room. On another level, the new area became a useful semi-covered courtyard for storing firewood and concealing the boiler.

This vertical elevation is the leitmotif of the renovation: it introduces external, changing elements such as light, rain, wind and clouds, while also structuring the interior by creating a fluid and dynamic relationship between different living spaces.

On the ground floor, there is a kitchen-dining room with a fireplace for cooking. The left-hand wall includes a large cupboard made of wooden slats, which blurs the perimeter and acts as an adjustable light source. The paving is made from stone, which makes it highly resistant and able to stand up to intensive use. A stainless steel worktop with a lacquered glass front section, and the cylinder-shaped extractor fan that was custom-designed by Walter Hegnauer, complete the kitchen area.

On the first floor, the living area and study occupy a single space that includes a chimney. A linear maplewood bench with drawers and adjustable lighting from below is one of the notable furniture elements, together with the worktable that has a glazed mandarin-color top and rests on the partition wall-railing, jutting out over the double space to the ground floor. Here the flooring is made from bamboo.

The bathroom was designed to act as a darkroom for photography, so light boxes were used to allow images and colors to be changed in order to provide different environments.

On the second floor there is the bedroom, dressing room and WC, forming a single space that extends out to the terrace through a large window. The glass skylight and floor connect visually with the first floor and the sky.

The shower is a transparent glass cylinder, designed to act as a lampshade at night. The interior flooring is bamboo, and the outdoor platform is made from copper pine. The lacquered DM dressing room is placed behind a partition wall that also functions as the headboard.

The project integrates furniture and lighting with the idea of creating a simple and serene environment.

On the ground floor, there is a kitchen-dining room with a fireplace for cooking. The left-hand wall includes a large cupboard made of wooden slats, which blurs the perimeter and acts as an adjustable light source. The paving is made from stone, which makes it highly resistant and able to stand up to intensive use.

This vertical elevation is the leitmotif of the renovation: it introduces external, changing elements such as light, rain, wind and clouds, while also structuring the interior by creating a fluid and dynamic relationship between different living spaces.

Rüdiger Lainer

Penthouse Seilergasse

Vienna, Austria

Photographs: *Margherita Spiluttini*

A two-storey penthouse conceived as a "House on a House": a transparent entity replacing the existing pitched roof, on one of the first reinforced-concrete apartment houses in the city. The site is two minutes from the Stephansdom in the 1st district of Vienna's "holy" area, and has a direct visual link to the cathedral.

The façade of the existing building, built in 1911, imitates the historical character of the street: however, it is in fact a concrete building, thereby necessitating a thoroughly modern approach to the penthouse.

Floor plates of exposed profile-metal sheets, as well as a reinforced concrete deck, supported by a steel frame, create a flexible, open space, and allow multiple room configurations. The external skin of stainless steel uprights with clamped glass sheets creates a link between the internal wooden floor and the external planted terraces, with unobstructed views of the dense urban fabric outside.

The project is programed to contain 5 entities, the combination and configurations of which allow the different zones to be used as either living or office space, or a combination of both. Furthermore, laminated glass is used as a roof in the central area to create the conceptual "rift" in the plan, thereby conveniently dividing the intervention into two distinct but interrelated zones. Interior space extends horizontally into the city and vertically to the sky.

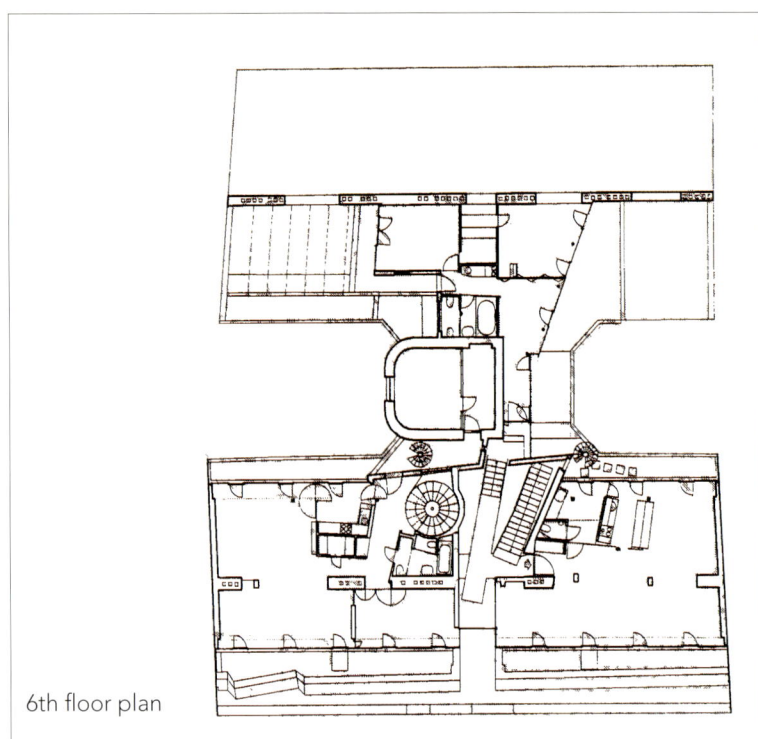

6th floor plan

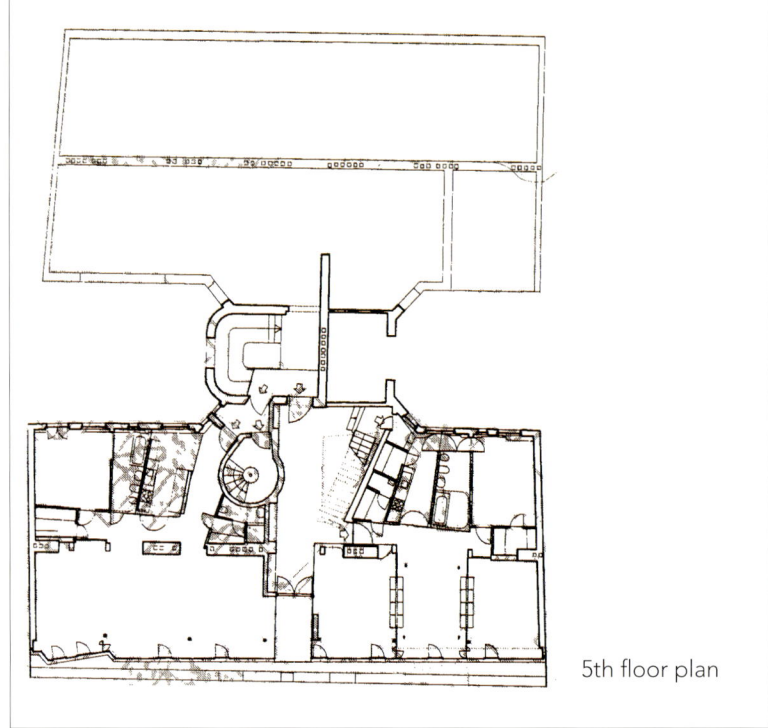

5th floor plan

Section 2-2

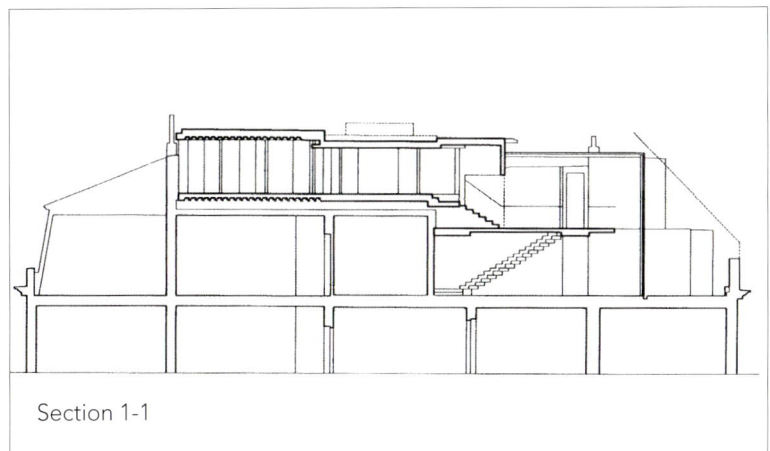

Section 1-1

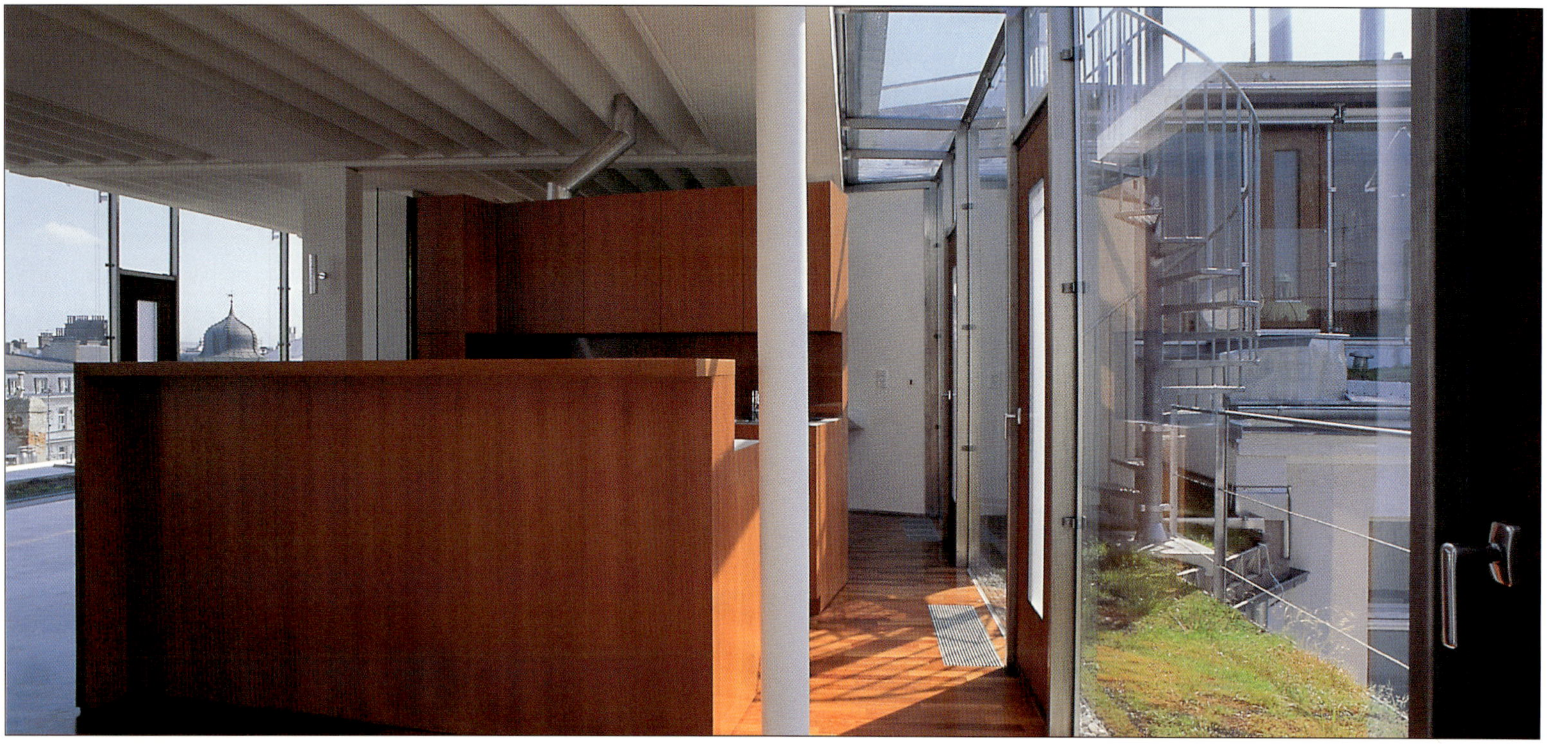

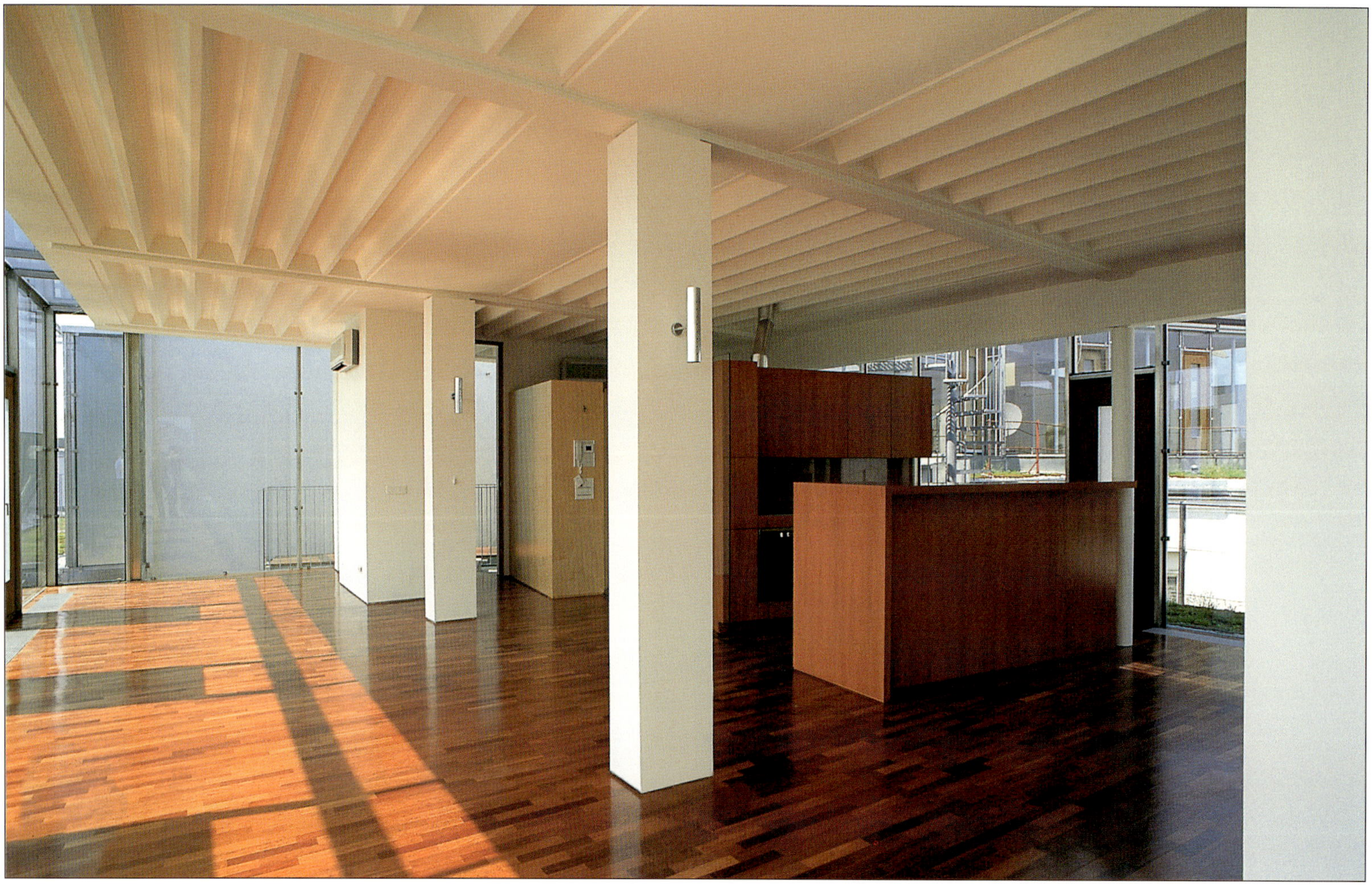

Uras + Dilekci Architects

Misir loft

Istanbul, Turkey
Photographs: *Ali Bekman*

The Misir building, designed by Armenian architect Hovsep Aznavour in 1910, is located in the central Beyoglu district of Istanbul. The brief por this project was to create a 280 sqm (3,000 sqft) loft on the second floor of this building to serve as a second city home for a couple who enjoy entertaining.

With this in mind, the architects focused on retaining some of the original character of the space, while transforming it into a highly original, modern and flexible apartment.

Near the entrance and in the kitchen, the floor was lifted to create a counter which can be used for dining and working. Changing color strips of light were embedded in the floor below, and can be programmed to create different moods. The lighting in most of the house, including a custom made chandelier, is made from simple black electrical wire and hanging lightbulbs, and strategically placed mirrored surfaces create different effects.

The original brick and structural timber was exposed in places in order to keep the original flavor of the building, and the plaster ceilings were burnt with a torch to create an organic texture.

In the bedroom, which can be completely closed off from the rest of the apartment by a 360 degree black velvet curtain, one of the walls is made of glass and the other is a remote controlled, red PVC lacquared door.

In the bathroom, oval marble and steel pieces were cut out and placed into the screed to create a terrazzo-like textured floor, and the cabinets were custom made using bamboo verneer.

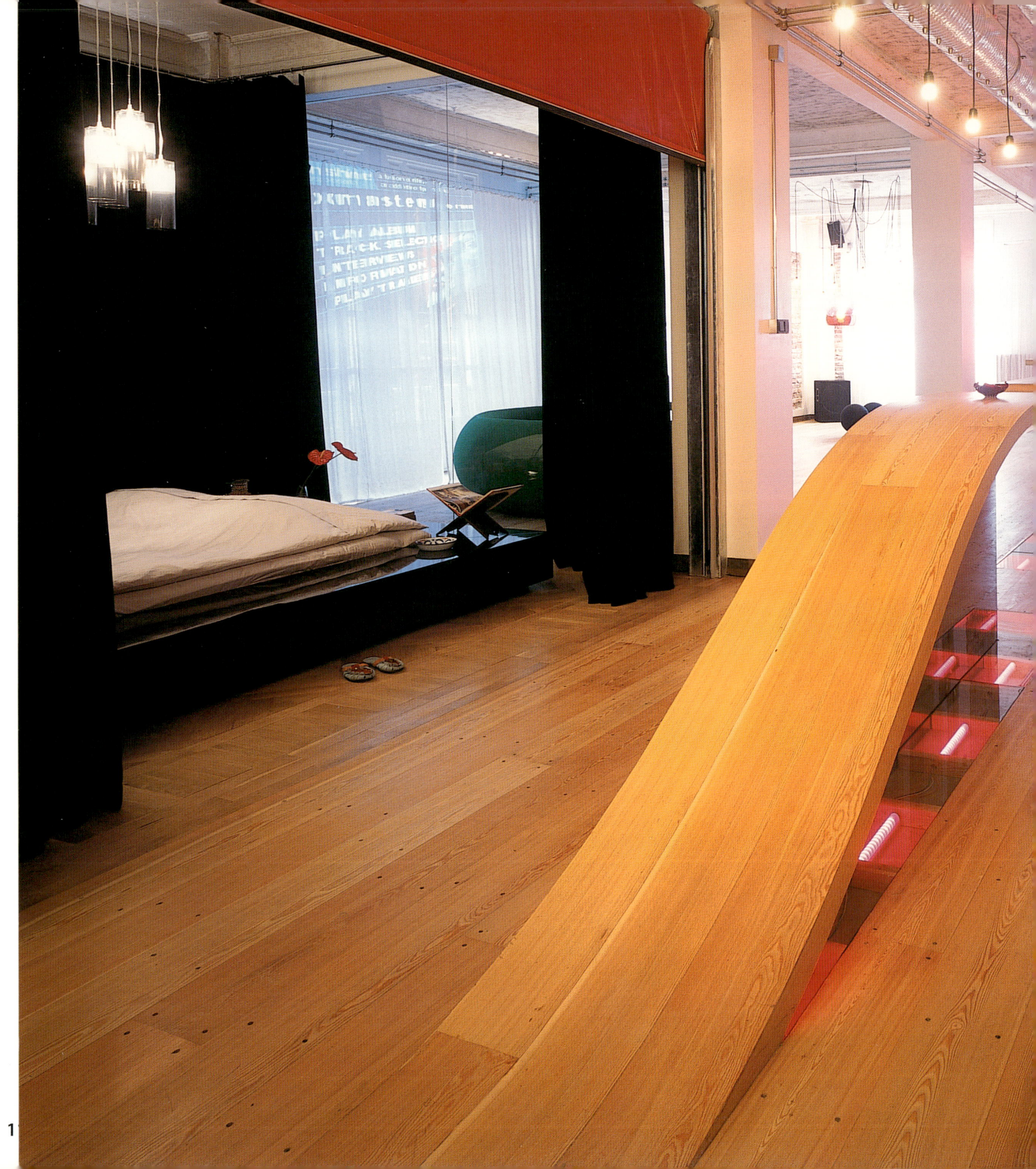

The bedroom is the only "closed" space in the apartment, and can be isolated from the rest of the space using black velvet curtains. A green circular dining booth was designed by Atelier Derin to serve more formal dinners in the space, and a screen which slides down over the glass wall separating the bedroom and the dining area can be used for screening films.

Michèle & Miquel

Apartment in Virreina

Barcelona, Spain   Photographs: *Josep Lluis Roig, Michèle & Miquel*

The program called for the design of a home/office/events space in a defunct butcher's shop. Juggling the concepts of mass/void/light, the decision was made to open the two façades as much as possible, one end toward the public square and the other toward the courtyard. At the same time, all of the interior vertical elements that divided or obstructed the space were eliminated. Like in Japanese gardens, the design sought out the furthest horizon and then interiorized it: the horizontal horizon would be the wide plane of paving in the square, interrupted by the benches and trees, as well as the façades in the background. The vertical horizon would be made up of the leaves of the creeping vine winding its way up toward the light and sky of the courtyard. White was used liberally in order to keep the boundaries at a visual distance.

By simply installing a horizontal platform and lowering the floor in two-thirds of the available ground space, four zones in one were achieved, each with its own environment and views. One space, the light-filled day room, is the more extroverted of the four and is a natural extension of the paving in the square. Another space lies beneath the installed loft and has been set aside for the dining room and for domestic chores. A vertical space overlooking the courtyard serves as a sitting room that is particularly attractive at night, being lit by the courtyard lights. Finally, a loft suspended beneath the ceiling is an intimate space with views over the "horizontal" and "vertical" sitting rooms on either end.

A wood veneer covers the entire floor, folding up to form stairs, and pulling away from the wall and floor to create closets, shelves and benches. Here it pulls out to reveal a table, there a bed, where it also serves as a headboard. Wood panels slide open to provide access to the bathroom, which is splashed in neutral tones and diverse plastic forms forming benches, chairs and numerous mobile objects.

The curtains made from strips of translucent plastic, vestiges of the old butcher's shop, capture shapes, motion and colors. The narrow, repeated relief of each strip multiplies and refracts the lighting conditions. With all of the curtains closed, the space becomes private, more intimate. When open, with the table and bed stowed away, the space has a much more public character and can be used as a gallery or impromptu music venue. All intermediate solutions are possible.

A small acacia and six empelopsi cover the party wall in foliage. In the summer, a fine sheet of water some 10 cm deep covers the courtyard, reflecting the light and vegetation, while at the same time comprising an excellent natural cooling device for the apartment.

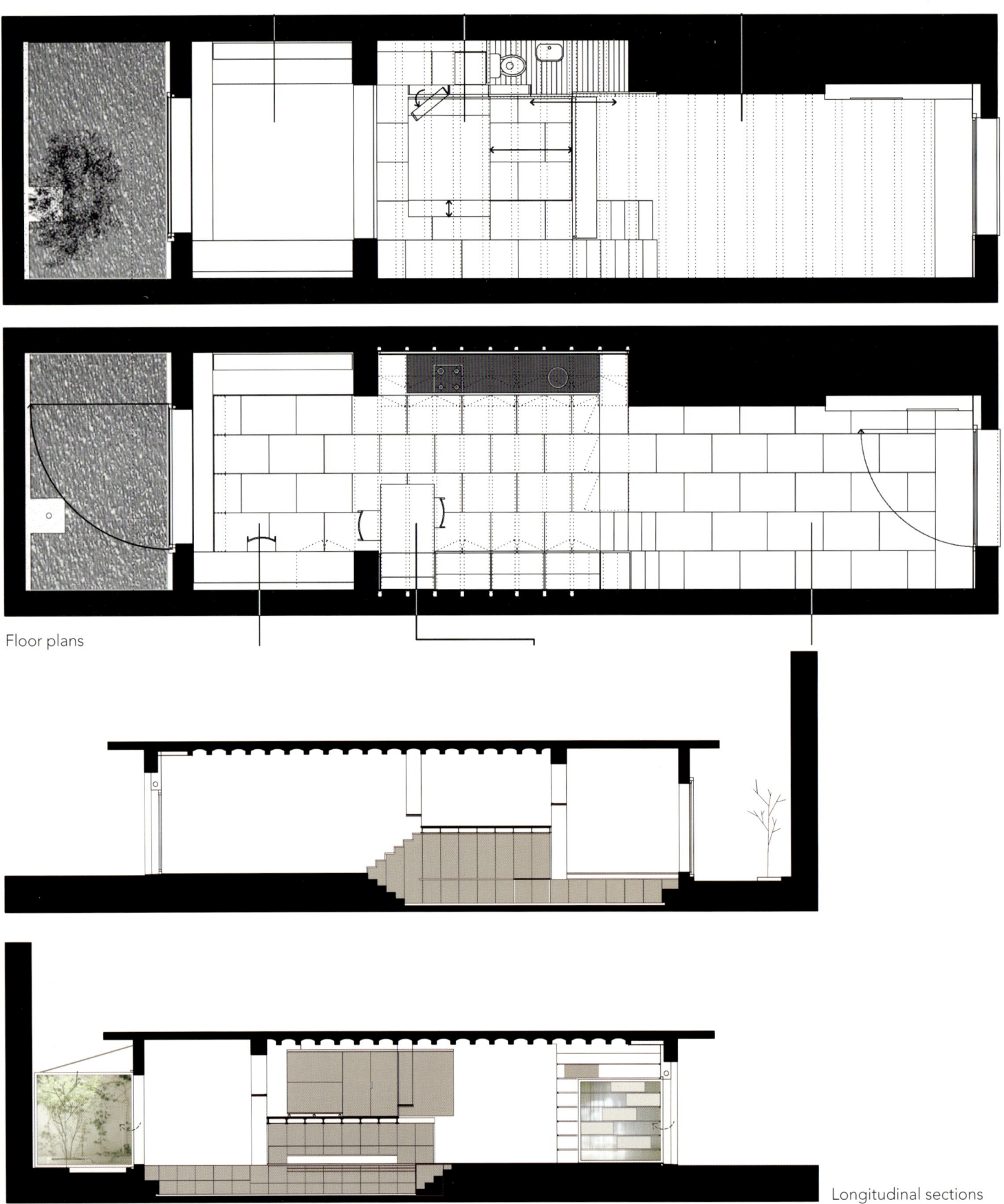

Floor plans

Longitudinal sections

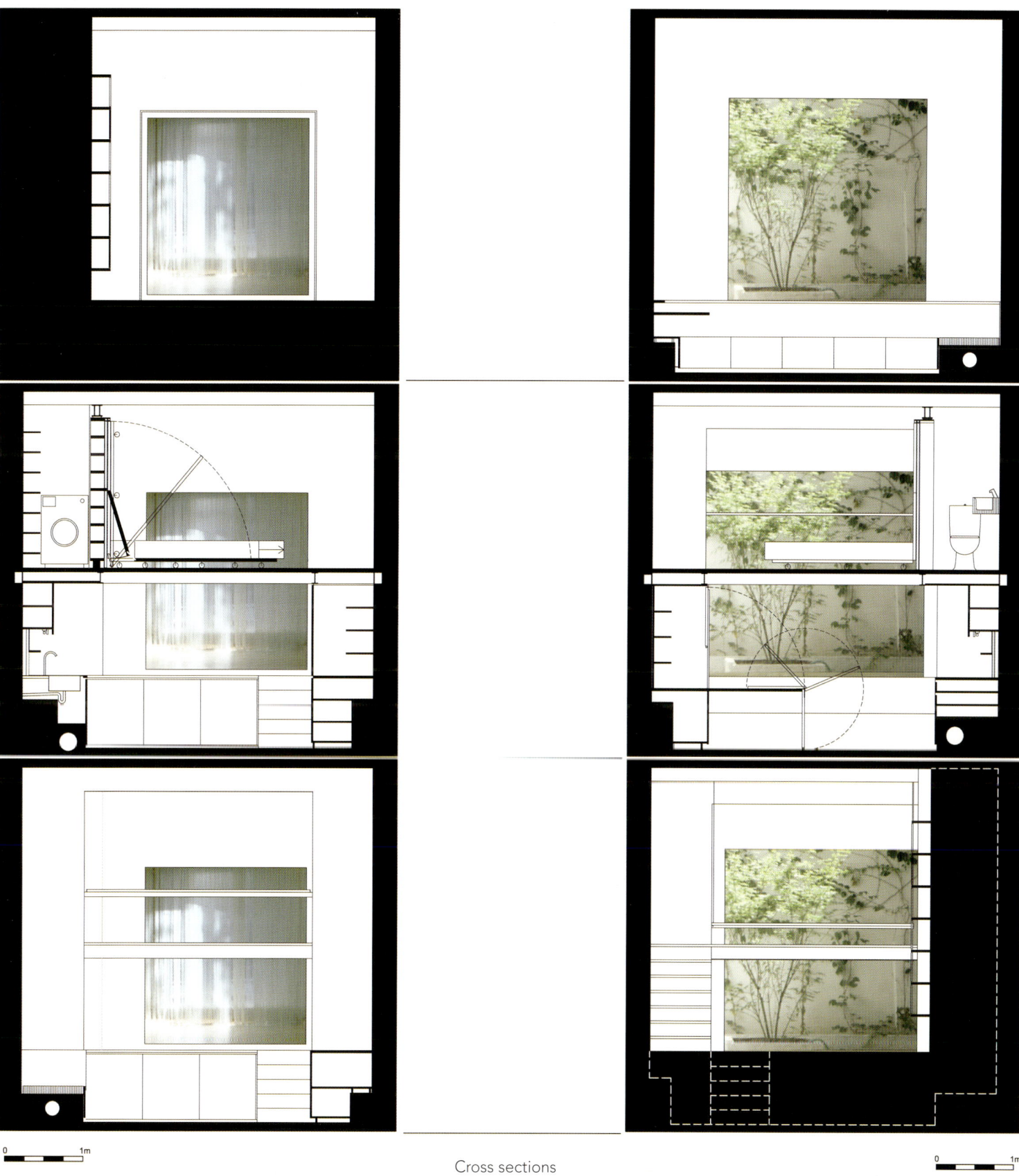

Cross sections

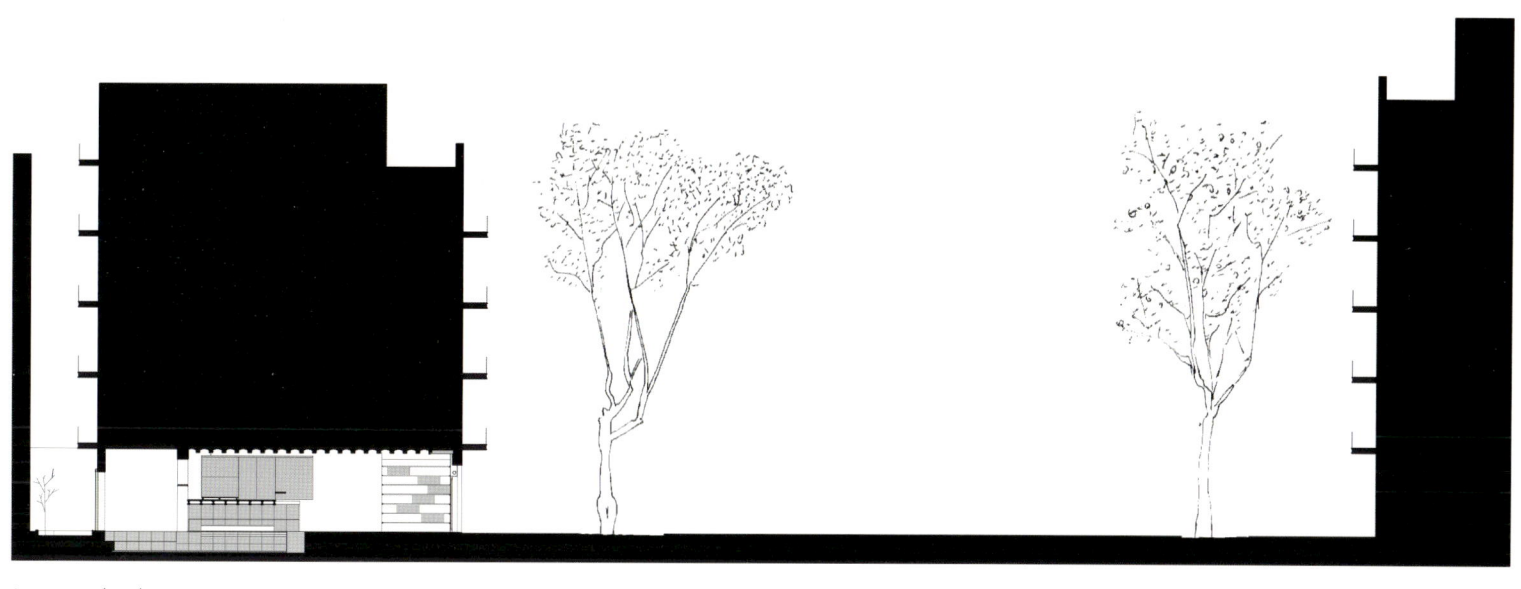

Longitudinal section

0 5m

MARC
(Michele Bonino, Subhash Mukerjee)

Apartment in Torino

Torino, Italy

Photographs: Beppe Giardino

A small, but unusually deep two-floor house is overshadowed among the massive buildings that have arisen all around it, in the Primo Governo period.

The building contains a 100 sqm (1,075 sqft) apartment that needs to make the most of a limited amount of natural light available. Accepting the lack of light within the central core of the building, this is kept dark on purpose, while its outer "shell" is designed to reflect the natural light powerfully throughout all the inner spaces.

The staircase that descends from the roof functions as a giant skylight. One of the showers is held by a structure suspended from the roof; direct natural light floods in through a skylight; the shower is provided with a translucent glass floor, which throws an agreeable light over the work desk below.

When the need is felt for a more direct contact with the environment, a 60 sqm (645 sqft) roof terrace is a convenient extra for sunbathing.

The central core of the building, which contains the bathrooms, drops down through the old house like an extraneous contemporary insertion, while the existing building conserves its typical characteristics, such as the plaster moldings, paneled carpentry and traditional French windows onto the balconies and the roof terrace.

Different flooring materials and patterns, from herringbone parquet floor, to tongue-and-groove planks, to stone or brick areas for different uses, define and outline the original distribution of the house, often contradicting the new distribution of interior spaces and creating a historical register of the building's earlier state.

The prevalent wall color is white, to reflect as much light as possible, but the areas that are only reached by diffuse indirect illumination are colored orange, creating a series of "hot" insertions that animate the potentially cold atmosphere. This gives certain functions, such as the kitchen cubicle, a separate character, although there is no real or structural segregation from the adjacent dining space.

The building's awkward and unusual volumes, plus the need to leave no storage space unused, required the special design and construction of much of the furniture and cabinetry.

Design team:
Michele Bonino, Luca Maletto,
Stefano Oletto
Contractor:
Make it snc, Torino
Furniture realization:
Marco Masoero, Lessolo
Surface area:
100 sqm + 60 sqm terrace

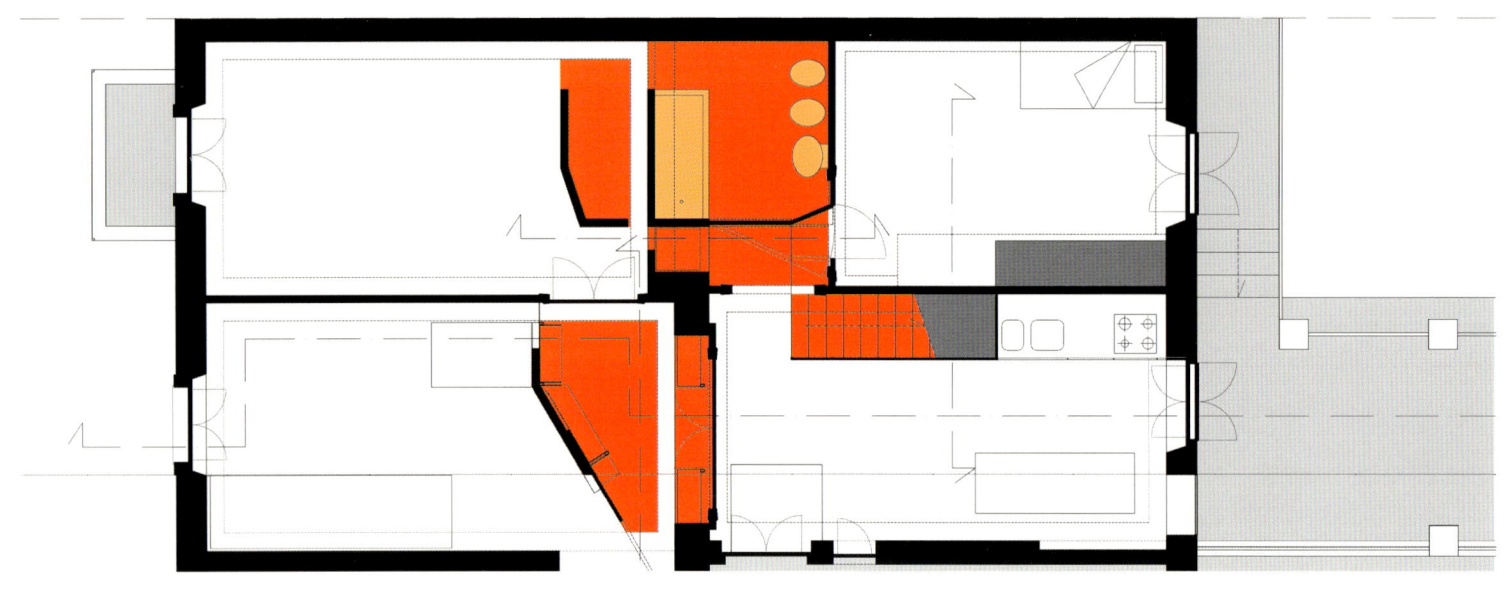

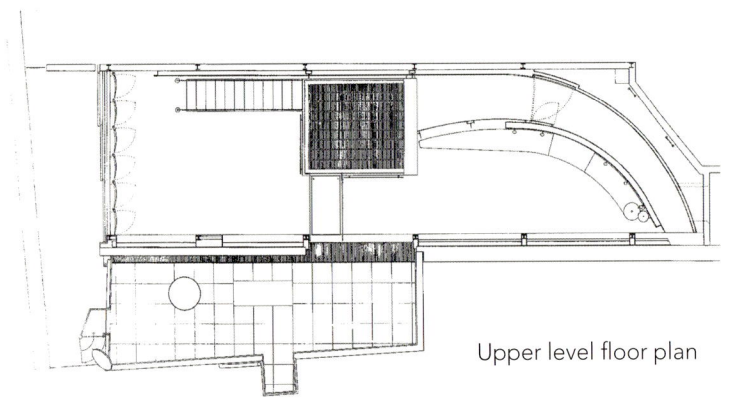

Upper level floor plan

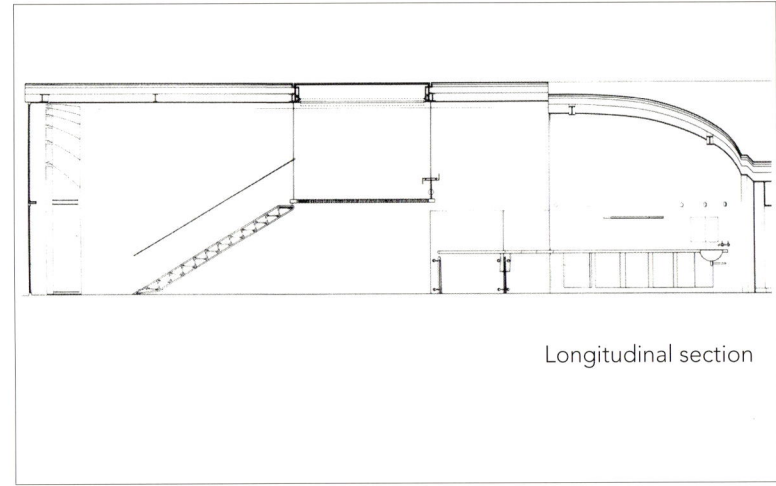

Longitudinal section

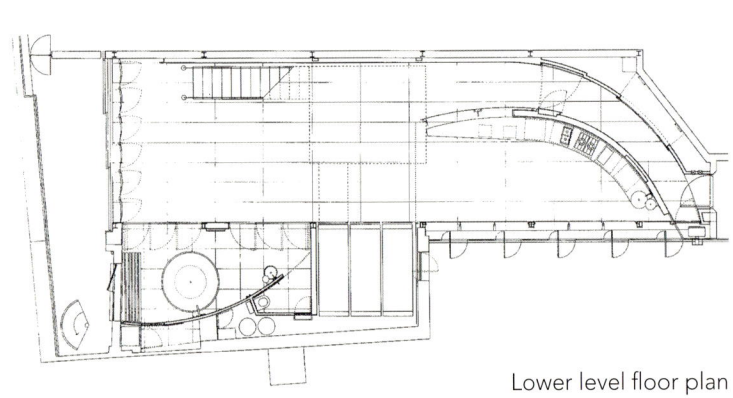

Lower level floor plan

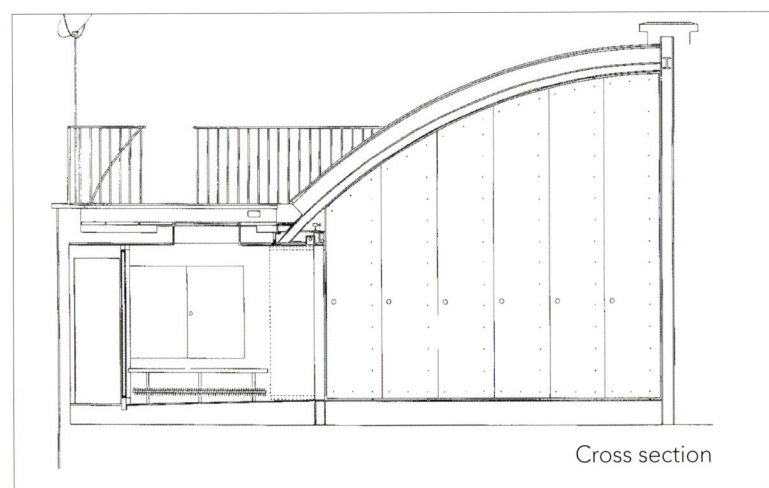

Cross section

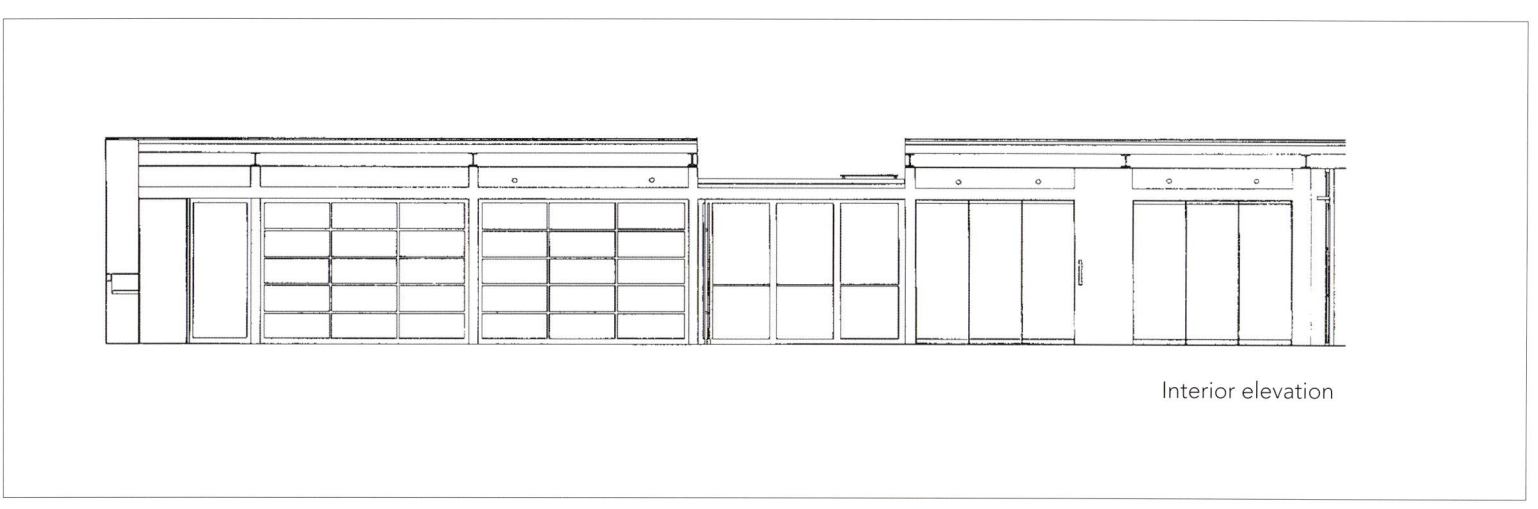

Interior elevation

The form of the existing shell was a very particular half arch with a small side space. A potential roof terrace was separated from the main volume and accessed via a lower terrace and a spiral staircase.

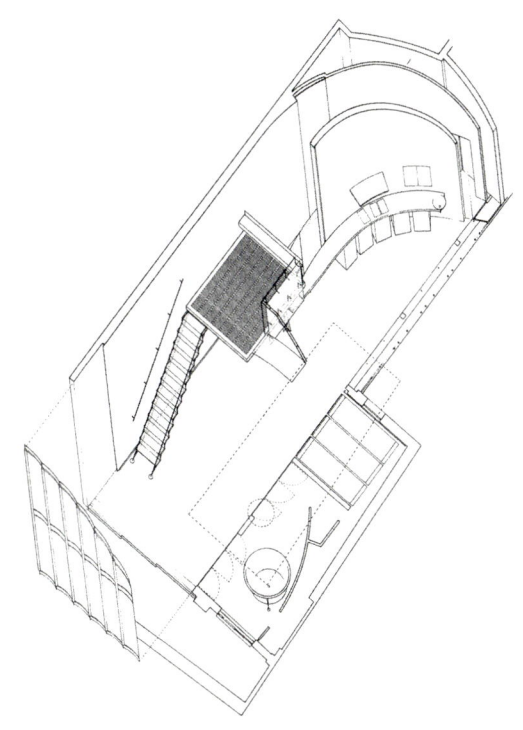

Manel Torres (IN Decoración)

Loft in an old factory

Palau de Plegamans, Spain

Photographs: José Luis Hausmann

This compact 70 sqm (750 sqft) loft is located in one corner of an old factory which continues working today as an industrial warehouse to store building materials. One end of the building was refurbished to create this home which is cut off from the day to day business of the warehouse, having its own entrance which gives directly on to the street.

The refurbishment has produced a result which is a far cry from the original use of this building: what was an old factory is now a young, dynamic space where day to day living is made easy and comfortable, two basic requirements for today's pace of life.

As the given space is quite narrow it was decided from the start to arrange an obstacle-free distribution, so the kitchen and dining room become one unique space, resolved according to the color scheme in the dining room area. This stems predominantly from the circular rug in which the black background acts as a base for a series of smaller circles laid out in concentric rings of different colors. These colors determined the color scheme of the different objects in the room.

In the center of the space is the dining room, consisting of a rugged-looking wenge stained oak table with straight geometric lines and surrounded by four chairs upholstered in red. This merges with the kitchen where the work surface is laid out in an L-shape, convenient for serving guests at the dining table as well as organizing the flow through the apartment.

The bedroom and the bathroom are the only rooms hidden from sight. The bedroom is situated on a mezzanine floor which was built in the ample space provided by the height of the loft. A staircase constructed on one of the side walls leads up to this area; its open stairs don't break up the visual flow through the space. The attic-like ceiling in the bedroom lends it a warm, intimate feel. The low-lying bed fills the center of the mezzanine on an oak base.

The mezzanine itself is made of pine which has been stained walnut color and its design is simple and straightforward: the low bed in the middle, a wardrobe and strong, brightly colored bed coverings, echoing the style of the main floor below.

The bathroom is tucked in behind the kitchen. Here it was decided to introduce a slightly rustic feel, with a terracotta flooring, exposed beams and a sponged treatment in earthy tones on the walls . The state-of-the-art bathroom fittings provide the perfect contrast with their innovative modern lines.

This modern refurbishment has a young person's dynamic style. It has no pretensions but on the contrary is all about making life easier and providing its users with a space that satisfies their needs and provides all the necessary services.

Jahn Associates Architects

Grant House

Sydney, Australia   Photographs: *Brett Boardman and Ghaham Jahn*

The Grant House is located in the back streets of inner city Sydney, where rows of terrace houses are separated by small-scale industrial warehouses. The original exterior brick walls have been retained, acting as an apron to the layers of timber and steel that make up the new façade. Two of the pine trusses from the original building have been reused in the first floor living area. Entry is through the outer brick skin and is marked by a galvanized steel lintel plate.

The idea of the home as a sanctuary in the city is central to the design, defining the organization of the spaces within the building and their materiality. The house is stacked to the south, enabling a ground-level, north-facing courtyard to emerge. Situated within this tranquil and contemplative space is the timber-clad studio, designed as a workspace and meditation room. It is placed to take advantage of the courtyard's water feature as well as providing a visual and physical link with the street entry.

The interior spaces are calm and protective, wrapping around the courtyard, and designed to accommodate an extensive art collection, as well as its future expansion.

The interior play of spatial relationships and materials in conjunction with the folded planes of the exterior, which wrap the surface of the building through its successive skins, unite to both embrace the life of the city and shun it. In addition, the simplicity of the new volume brings a new dimension to the street and encourages interaction through the natural aging of the materials, the original brick shell acting as the catalyst for an inventive and human response to the experience of living in the city.

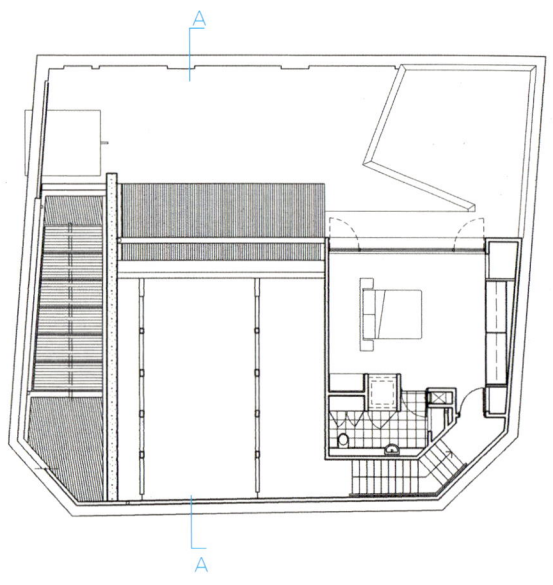

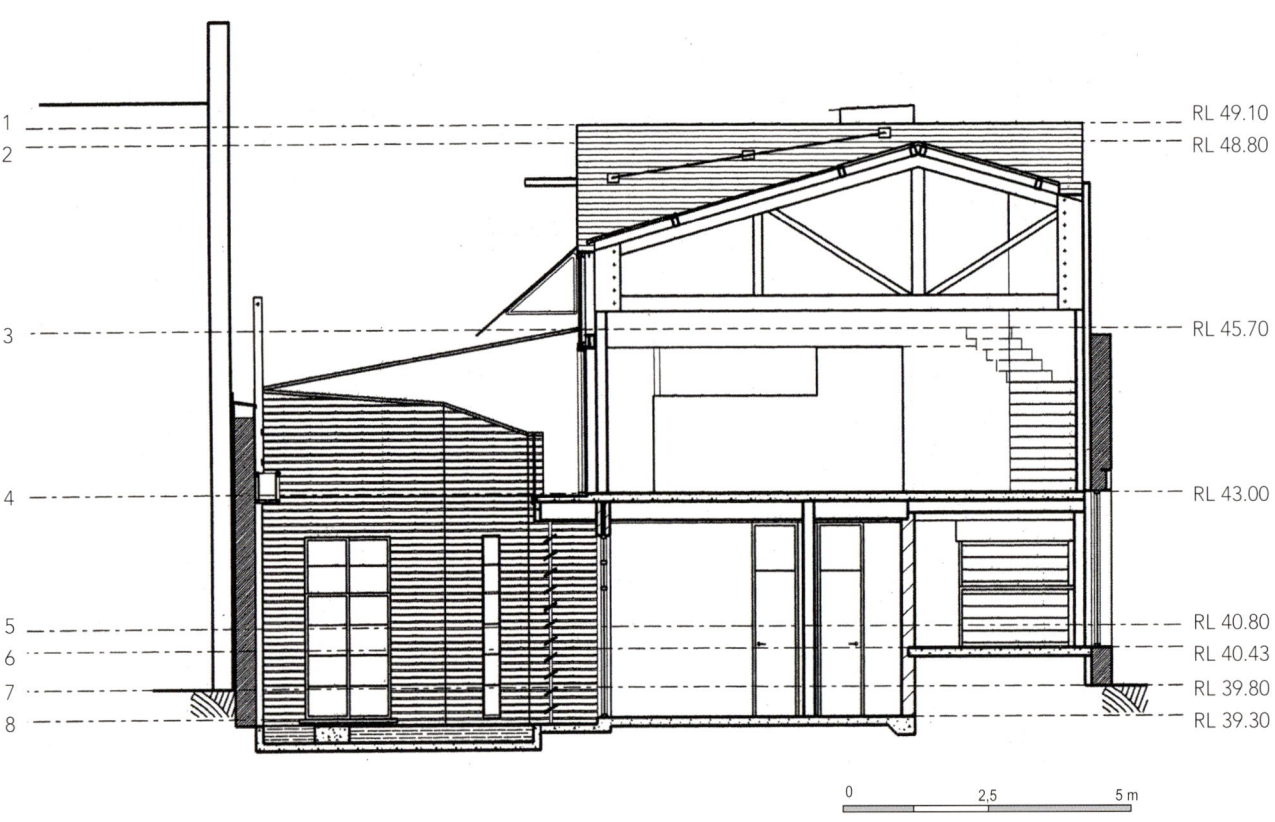

AA Section

1. Second floor roof
2. First floor roof
3. Second floor master bedroom
4. First floor living
5. Raper St. entry
6. Ground floor
7. Studio floor
8. Existing ground floor

RL 49.10
RL 48.80
RL 45.70
RL 43.00
RL 40.80
RL 40.43
RL 39.80
RL 39.30

0   2,5   5 m

MARC
(Michele Bonino, Subhash Mukerjee)

House in Via Barbaroux

Turin, Italy

Photographs: *Beppe Giardino*

On the first site visit with the client in this attic in the historical part of Turin, the contractor fell downstairs as the floor cracked under his feet. It became quite clear that reinforcement was necessary, but this would absorb almost the whole budget, not leaving any resources for other designed work.

The architects decided that the whole project should be the reinforcement itself. Taking advantage of the roof section, the reinforcement is split into two halves, with one part raised 75 cm. The architects exploited the difference in levels, cutting and deforming the edge of the new slab to generate much of the needed furniture: the floor of the bathroom becomes the kitchen, the living room becomes the dining table, the changing space becomes the bed.

The rest are little things - two glazed walls separate the kitchen from the bathroom and the bed from the entrance. The toilet and storage area are hidden behind movable zinc surfaces.

The difference in level allows a view of the nearby church dome from the high roof windows, and makes it possible to recess a tub into the bathroom floor, with a view onto the kitchen sink.

The floor is olive wood from Calabria, the client's homeland, but only on the upper level. To emphasize the difference between the two levels, the lower part is treated with a more budget-friendly waxed concrete.

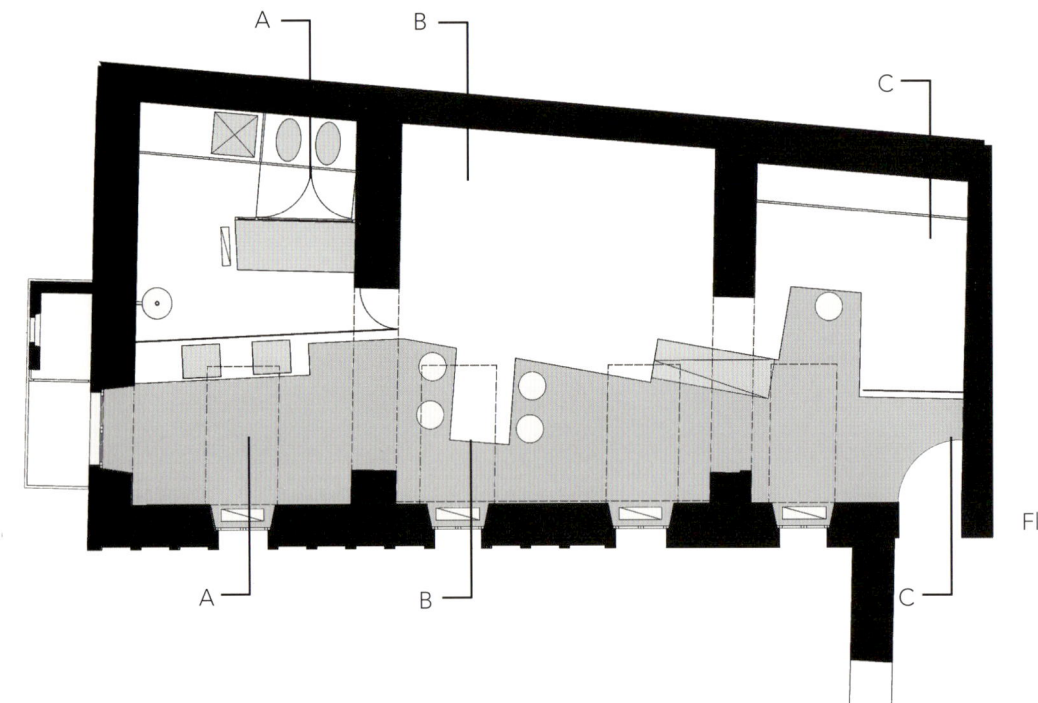

Floor plan

The architects decided that the whole project should be the reinforcement itself. Taking advantage of the roof section, the reinforcement is split into two halves, with one part raised 75 cm. They exploited the difference in levels, cutting and deforming the edge of the new slab to generate much of the needed furniture: the living room becomes the dining table, the changing space becomes the bed.

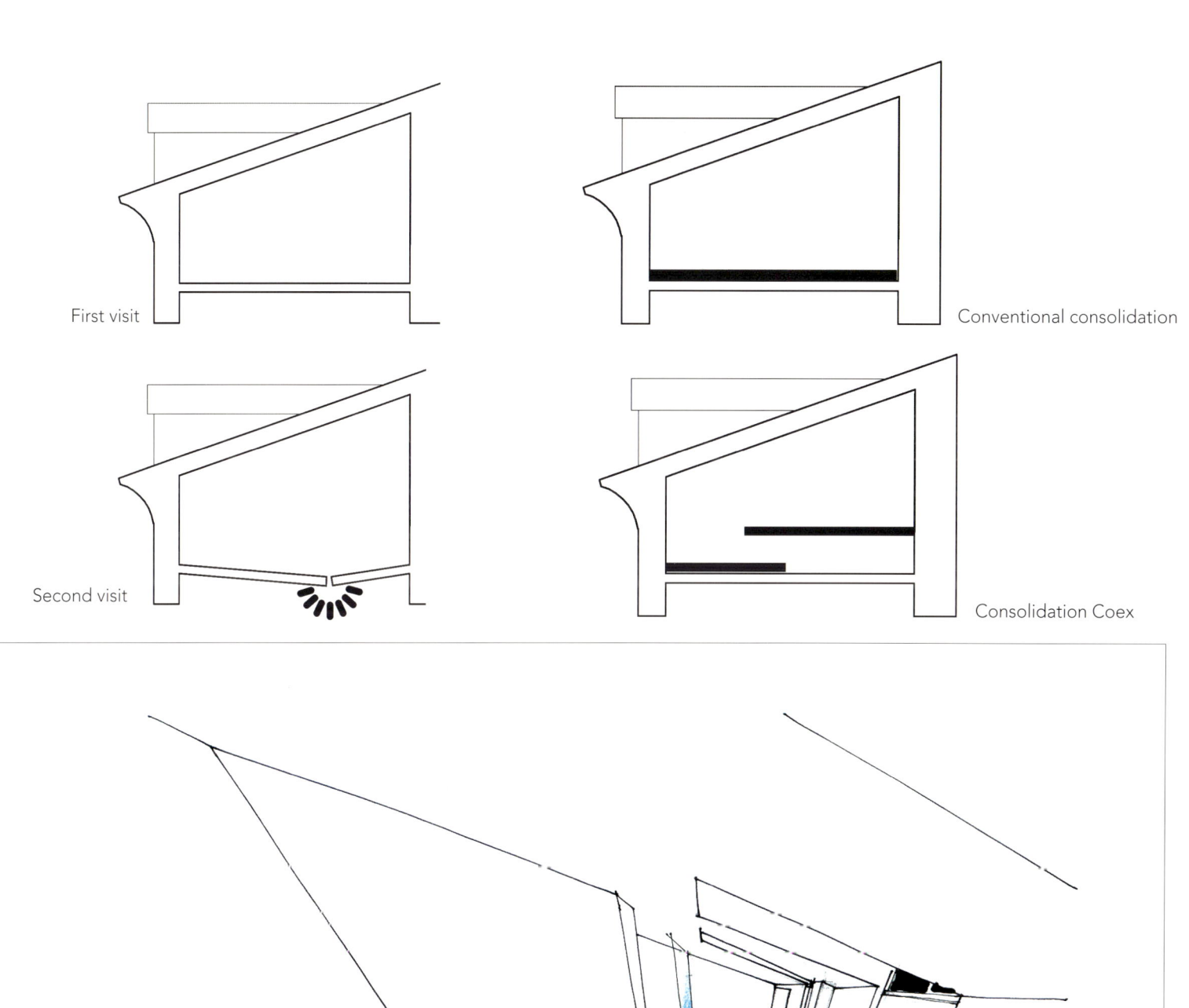

First visit

Second visit

Conventional consolidation

Consolidation Coex

Giovanni Scheibler

Loft Conversion in Zurich

Zurich, Switzerland

Photographs: *Alex Spichale*

In converting the loft space in this house, the aim was to create extra living space and to bring more light into the top floor flat, while respecting the characteristic turn-of-the-century outward appearance of the building. Ecological and economic factors were also taken into consideration. The architects created a central hall, lit from above by a new roof light set in the ridge of the mansard roof, 7m (23 ft) above the hall floor, There are no fittings that block the path of light: even the gallery floor is of clear glass. Light can also filter through translucent walls into the rooms bordering the hall. Sliding partition elements further help to create space and versatility, in contrast to the usual narrow confines of such flats. The materials used for the new hall are clearly legible against the existing building structure. Anthracite-colored metal, chrome steel and glass stand next to plaster walls and wood. The fine lines of the elements of the hall complement the theme of transparency. The girders supporting the gallery are pairs of tensioned RHS-sections, resting on brackets on the mansard structure.

The cable bracing eliminates any vibrations along the slim sections. The safety glass flooring sheets rest on a double layer of rubber to reduce noise. Chrome steel is used for the handrail and the horizontal cabling. The frame of the sliding partitions is of narrow square tube sections. Between the glazing layers is white glass lining welded to the panes at the side.

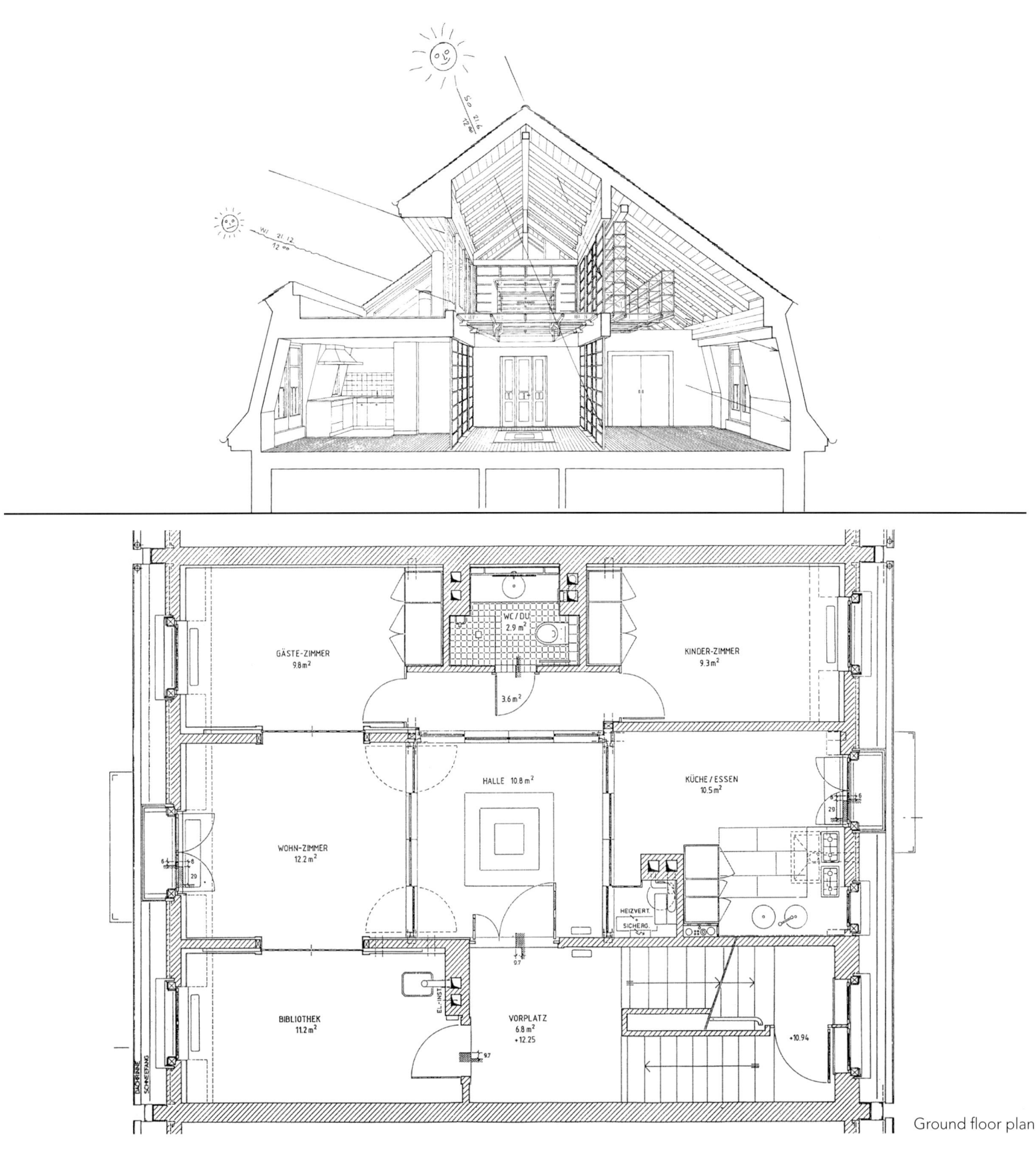

Ground floor plan

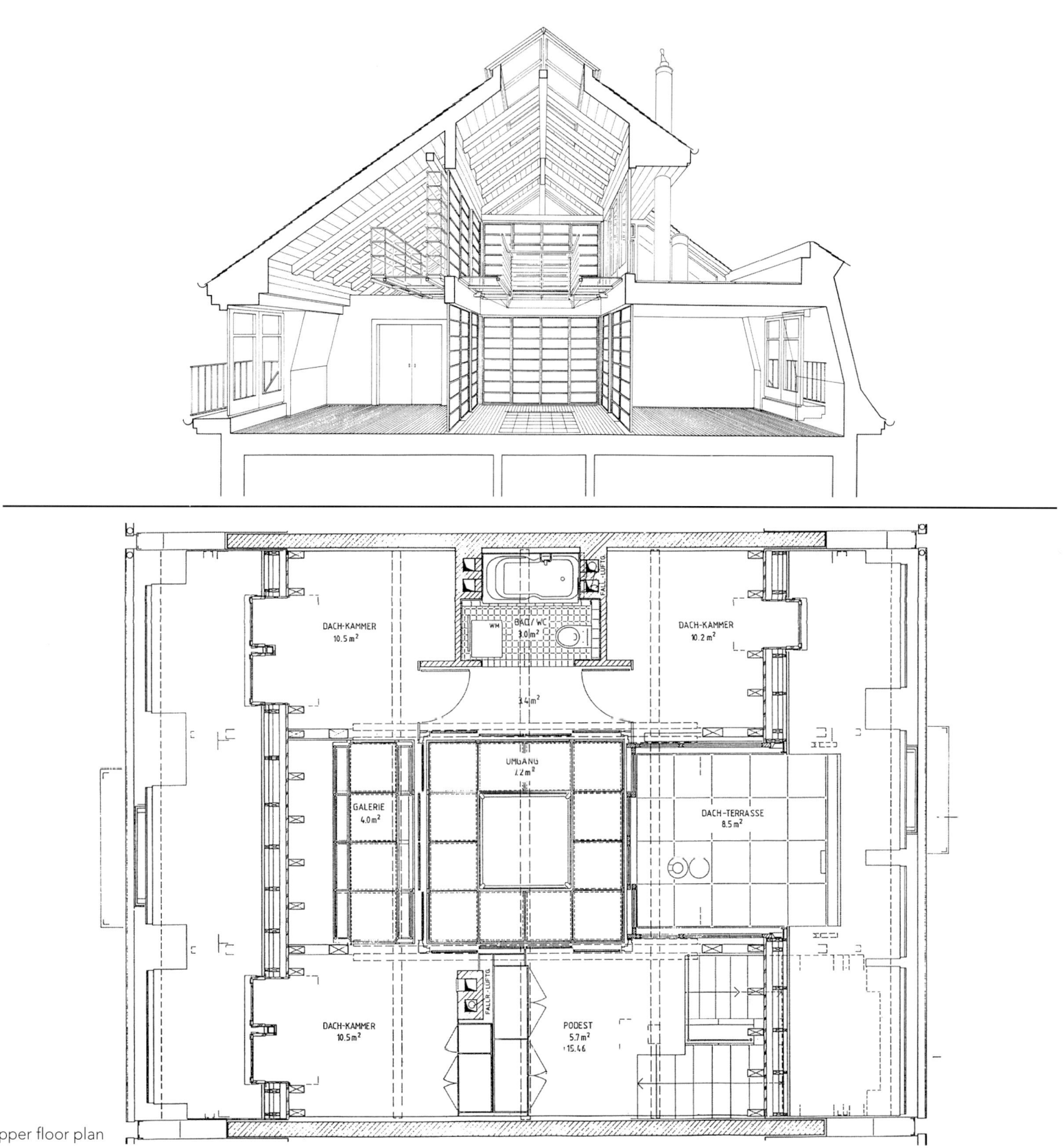

Upper floor plan

DACH-KAMMER
10.5 m²

BAD/WC
3.0 m²

WM

DACH-KAMMER
10.2 m²

1.4 m²

UMGANG
1.2 m²

GALERIE
4.0 m²

DACH-TERRASSE
8.5 m²

DACH-KAMMER
10.5 m²

FALLR-LUFTG

PODEST
5.7 m²
+15.46

The dwelling is organized around a central hall that is top-lit by a skylight. The light comes through the translucent glass panels into the rooms surrounding the central hall.

The materials used to build the central hall (steel and glass) are easily recognizable, contrasting with the existing structure of the building.

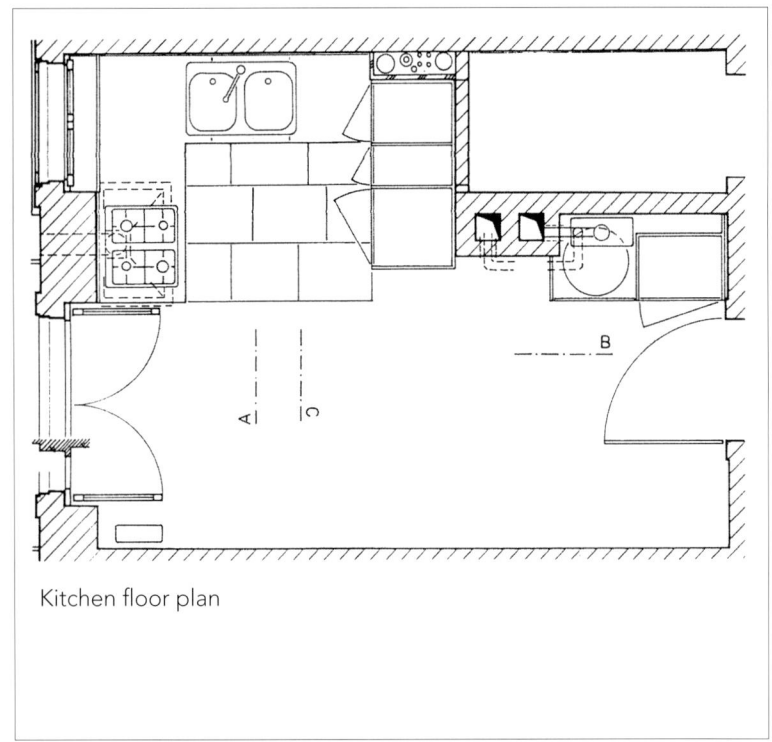

Kitchen floor plan

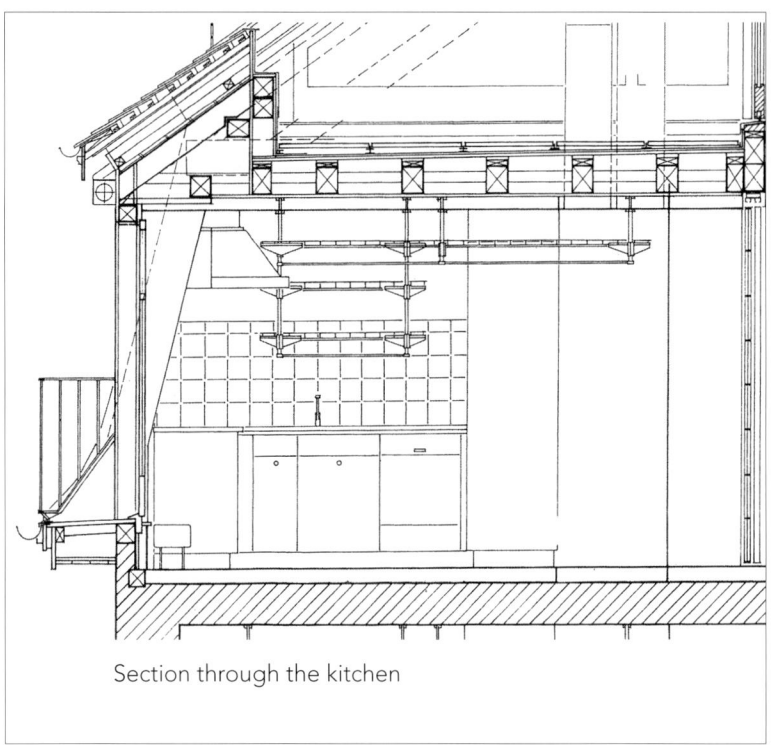

Section through the kitchen

The kitchen, reached directly through the central hall, has been completely redesigned to create two sharply differentiated spaces within it: the work area and a small eating area.

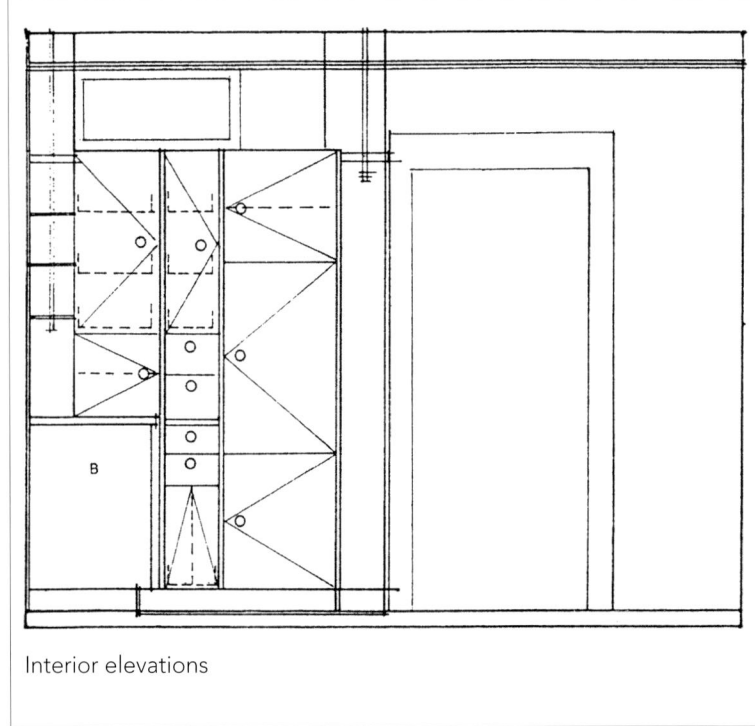

Interior elevations

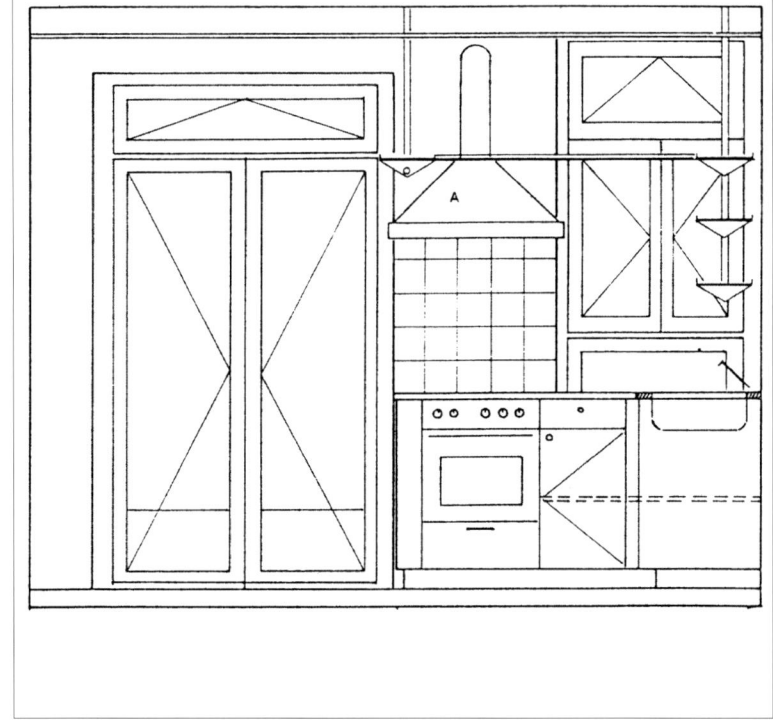

Floor plan

Daylight passes from the larger elements to the smaller ones: it enters from the large existing industrial windows, is modulated by the walls (of different heights) and filters through strategic openings into the walls.

The configuration of this installation is based on concentric, interpenetrating volumes; a series of progressively larger volumes inside one another (like a set of Russian dolls) all of which nest inside the shell of the existing industrial building.

Section

Illumination diagram

Marco Savorelli

Nicola's Home

Milan, Italy   Photographs: *Matteo Piazza*

The concept of the recovery of this old loft was the result of a close collaboration between the client and the designer. It developed from nothing: an abstract study of the functions and utilities of a "home system", developing from the renovation of a single, elementary space, in which old functions translate into new and simplified forms.

Intimate, operative spaces like bathroom, kitchen and wardrobe become monolithic volumes that, reduced into simple forms, are brought to the fore, creating new connections, and spatial relationships within the space.

In this project the historical memory of the site is subjected to thorough formal research. A well-balanced experiment with new spaces, which preserves the existing light quality. The result is a playful alternation of volumes and moods, a fluid exchange between the existing and the designed space. These are characteristics of a project which evolved from the intense dialogue between the architect and the client, aiming to achieve a minimalist aesthetic and at the same time volumetric and functional complexity. This is not a mere operation of interior decoration but the creation of volumes to be lived in and "live with" in a completely modern and innovative way. The space acquires both a jocose and a reflexive quality.

When entering this apartment the visual impact is instantaneous —a nearly flash-like perception of the space— which reveals the equilibrium between matter and light. The natural daylight traces delicate designs on the neat surfaces, shadows in perpetual movement creating a simple and primordial game of light and darkness.

Light was treated as an element of the construction whose function consists not only in lighting the dwelling but also in defining spaces inside it. In this sense, the strategic arrangement of top lighting gives the environment a very peculiar vertical dimension.

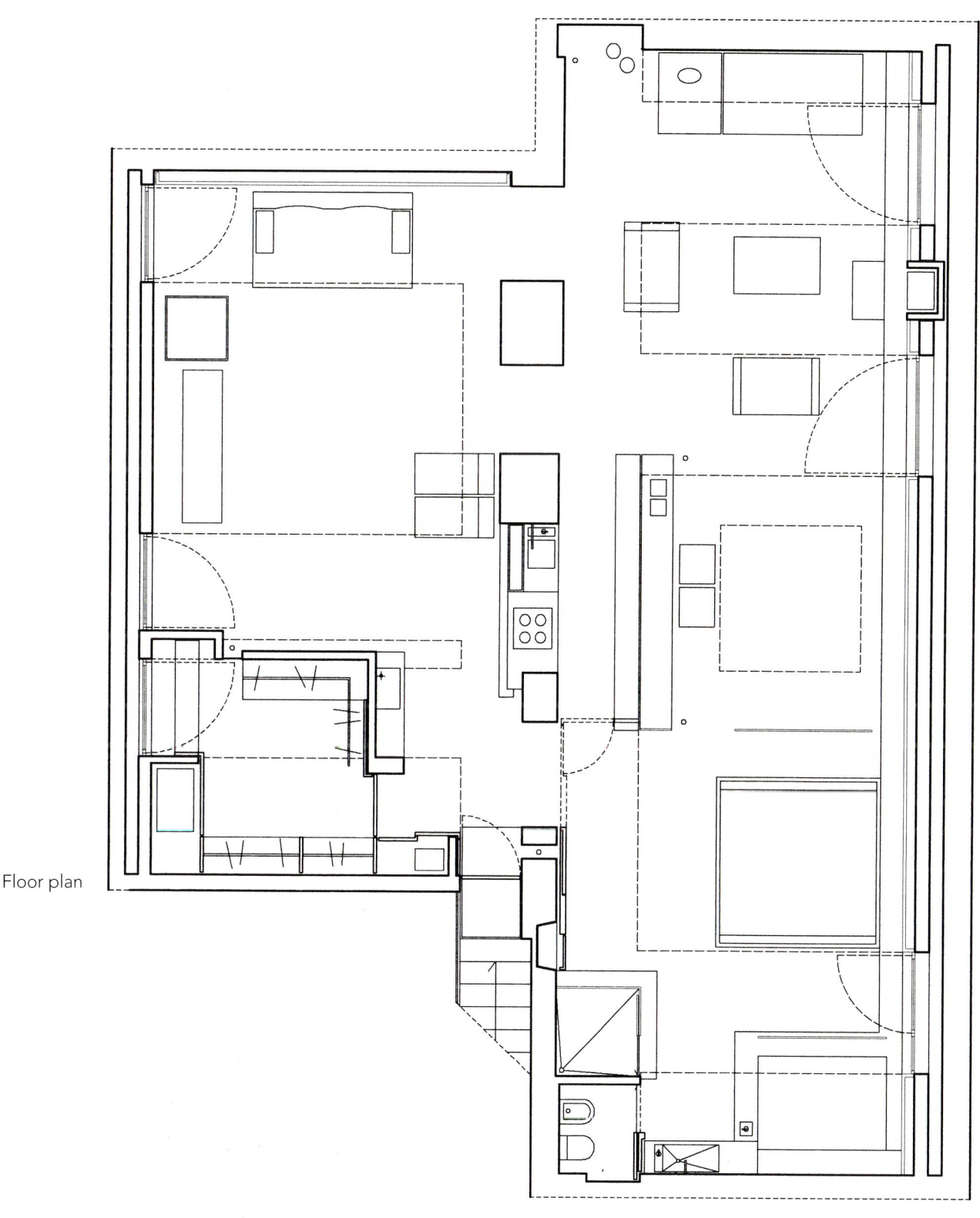

Floor plan

Otorino Berselli
& Cecilia Cassina

Restoration in Manerbio

Brescia, Italy

Photographs: *Alberto Piovano*

The house is set at the end of a closed alley inside the urban fabric of Manerbio. It had long been uninhabited and used exclusively for storing material and equipment by a local building firm.

The poor condition of the building and the inappropriate intervention on part of the portico in the sixties did not curb the imagination of the new owners, helped and stimulated by the considerable dimensions of the courtyard and the garden.

The fascinating volumes and the need for a clear solution to the disastrous intervention of the sixties suggested the idea of proposing a sequence of very recognizable architectural elements on both the façades. The original brickwork of the portico was revealed, the stone wall was cleaned and the wooden roof structure was rebuilt.

The three-storey house of the early 20th century was recovered with the typical structure of the epoch, a stone staircase leading to the different levels and dividing each floor into two rooms.

The central body from the sixties acts as an element of union and dates the intervention.

On the façade of the extremely rational cube, the evidence of the sequence of the pillars of the portico suggested the openings, which at night-time turn the interior spaces into a permeable and transparent box.

The organic nature of the materials and the almost obsessive repetition of a single color (floors, masonry, frames) conflict strongly with the wing of the laundry and kitchen in blood red and with the wall and the pillar of the living-room in exposed brickwork.

Views of the dining and kitchen area located on the lower level. These rooms are separated from the bedrooms by means of a fitted wardrobe that has curved metal doors with a matt finish and a swivelling door.

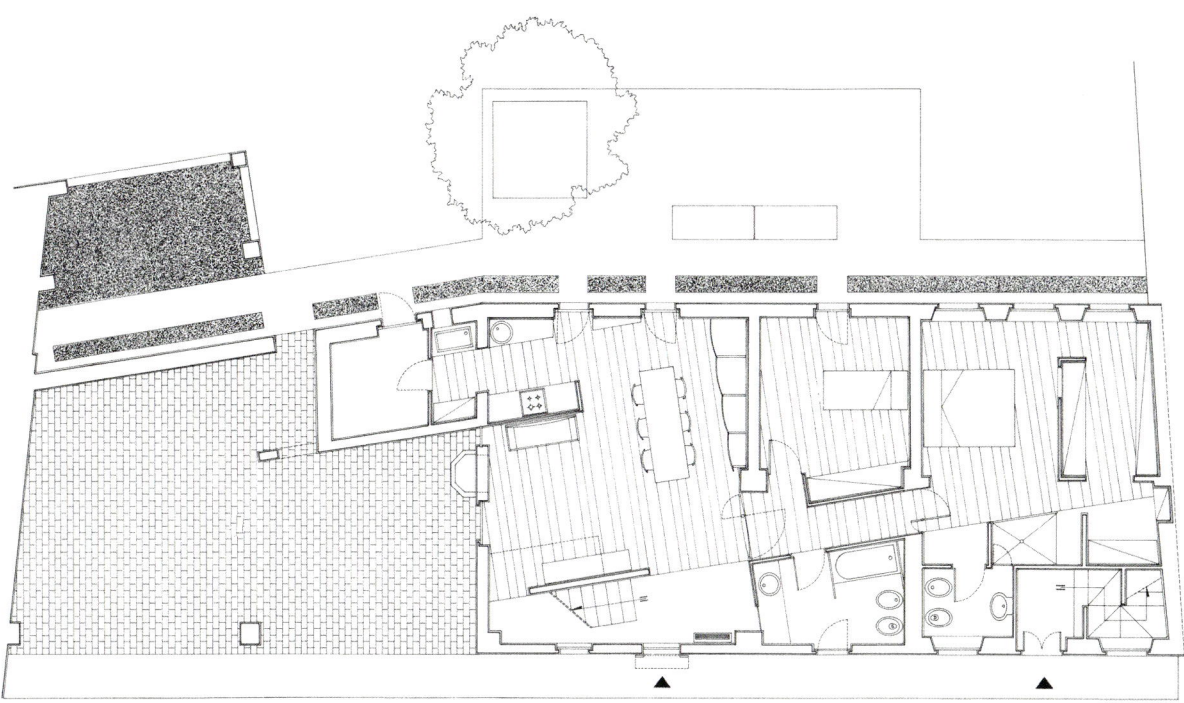

Ground floor plan

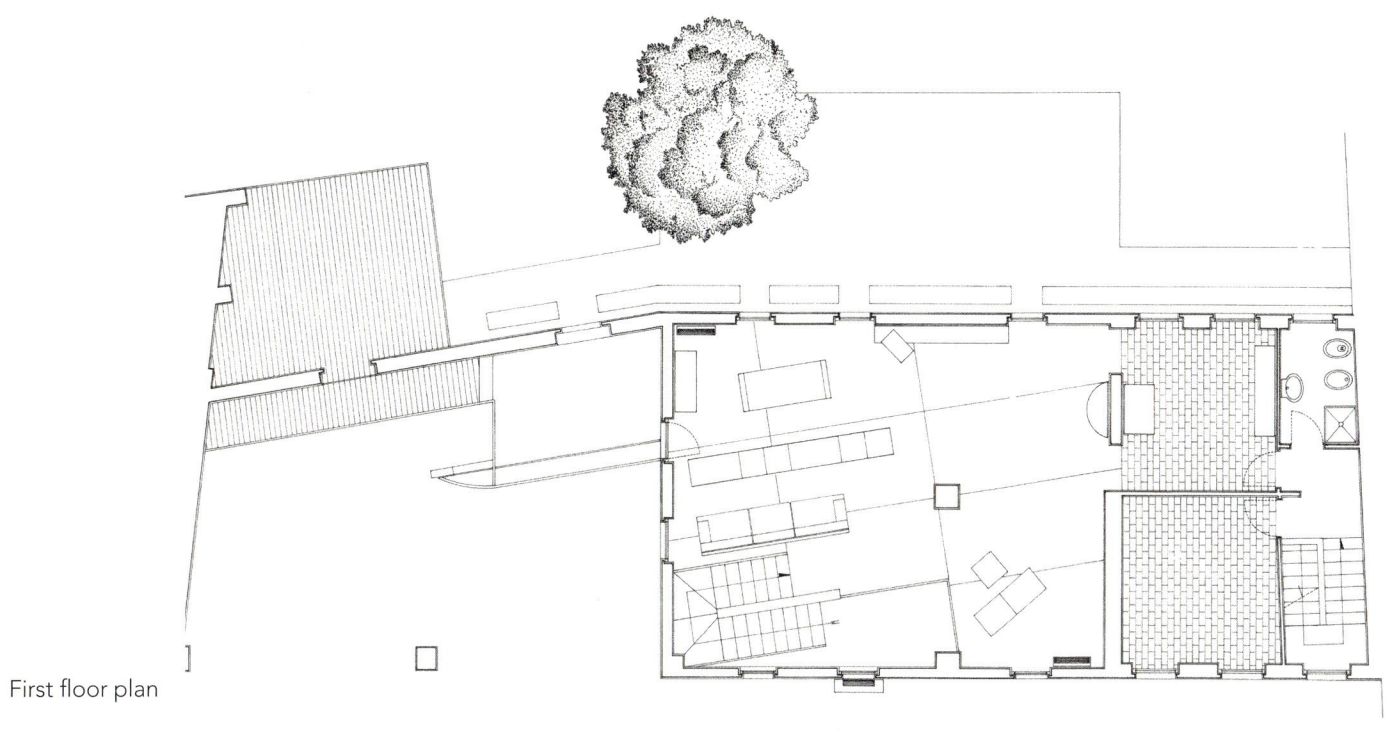

First floor plan

The two-flight staircase communicating the ground floor with the upper level is made of colored, reinforced cement to which thin steel bars have been attached as a handrail.

On this page are several views of the living area located on the upper floor. This communicates with the external terrace that looks onto the garden through a small balcony and a walkway anchored to the stone wall.

Guido Canali

Water Mill

Modena, Italy   Photographs: *Paola De Pietri, Alberto Muciaccia, Stefano Botti & Francesco Castagna*

The object of the reform work was an old building whose nucleus dates from 1558. It had been used for productive functions, such as carpentry, until 1896, when it was turned into a mill.

The base of the floor plan is a rectangle, the length of which runs alongside the canal and is divided into three areas. The main, northernmost space, where production took place, has a brick body and common pitched roof (corresponding to the mill). The grinding block is in the basement and, on the first floor, there is a large grain storage room with wooden trusses overhead.

Fortunately, the space which once housed the grinding block was in fairly good condition when it was acquired by the current owners - even the wooden scoops and millstones were still intact. The rest of the building, on the other hand, showed signs of having been subjected to a number of alterations.

The idea behind the project was to leave the millstone room intact, carefully restoring it while highlighting the theme of water, which had once been inseparably linked to the functioning of the mill. Thus, a glass bulkhead was installed, allowing observation of the interior of the waterworks and the scoop mechanism, both of which have been completely restored.

Adhering to client demands, a large apartment 1000 sqm (10,800 sqft) was created in the northern body and two independent apartments (104 sqm, 1,120 sqft and 130 sqm, 1,400 sqft) were installed in the southern volume, plus a service room of 40 sqm (430 sqft). The entrances to the master bedrooms are located above the canal at the middle of the building. Natural light filters through the skylight and is reflected off the swimming pool in the basement. Metal walkways cross the void of the foyer on the different floors, linking the building's two wings.

In the central volume, the garden has been turned into a kind of room/porch. Large sliding windows set behind a portion of the old wall allow this space to be either open or closed, depending on the season.

Throughout the scheme, particular attention was paid to the use of traditional materials for the indispensable integration of the old factory.

Structural calculation:
Francesco Canali
Collaborator:
Angela Cacopardo

The dining and living rooms are located on the top floor of the volume of the former granary. Two hand-worked skylights with stainless steel frames concealed between the existing beams illuminate the space.

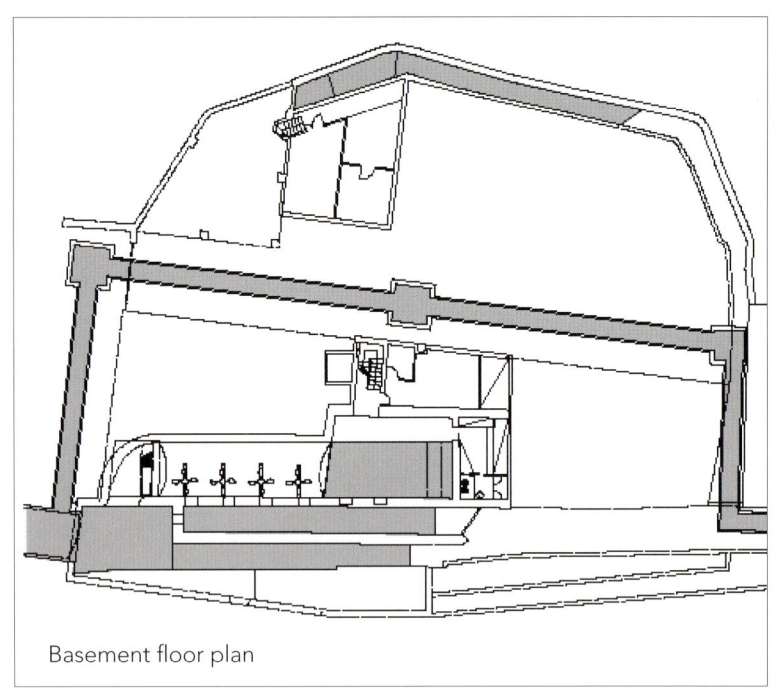

Basement floor plan

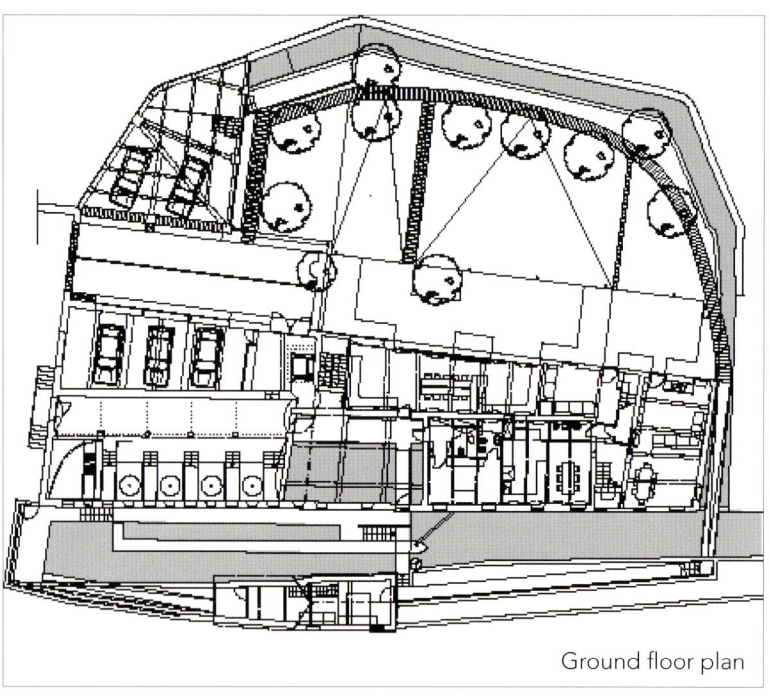

Ground floor plan

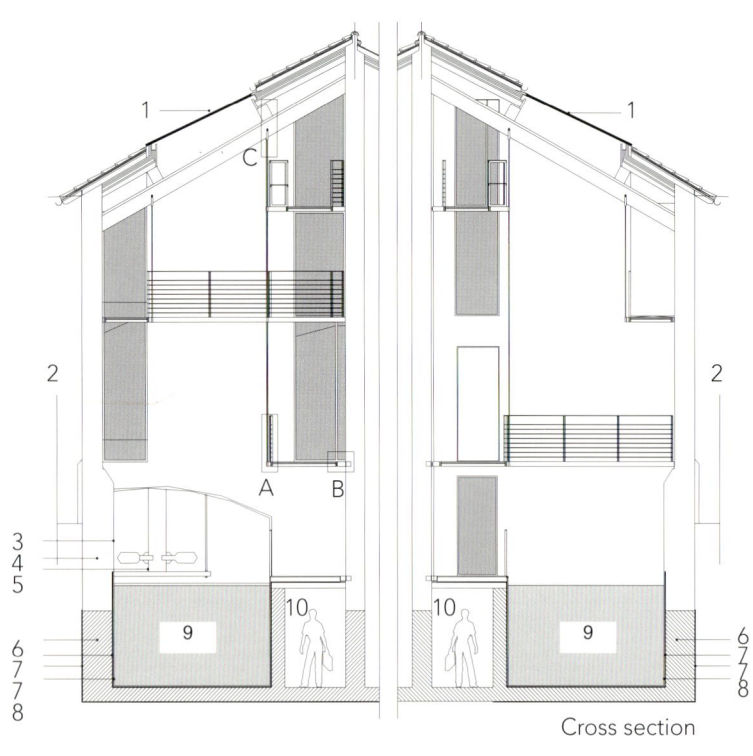

Cross section

2. Skylight
3. Barefaced brick
4. Old brick wall
5. Old mill blades
6. Reinforced concrete foundations
7. Insulating casing
8. Custom-cut ceramic plate cladding
9. Swimming pool
10. Swimming pool pump room

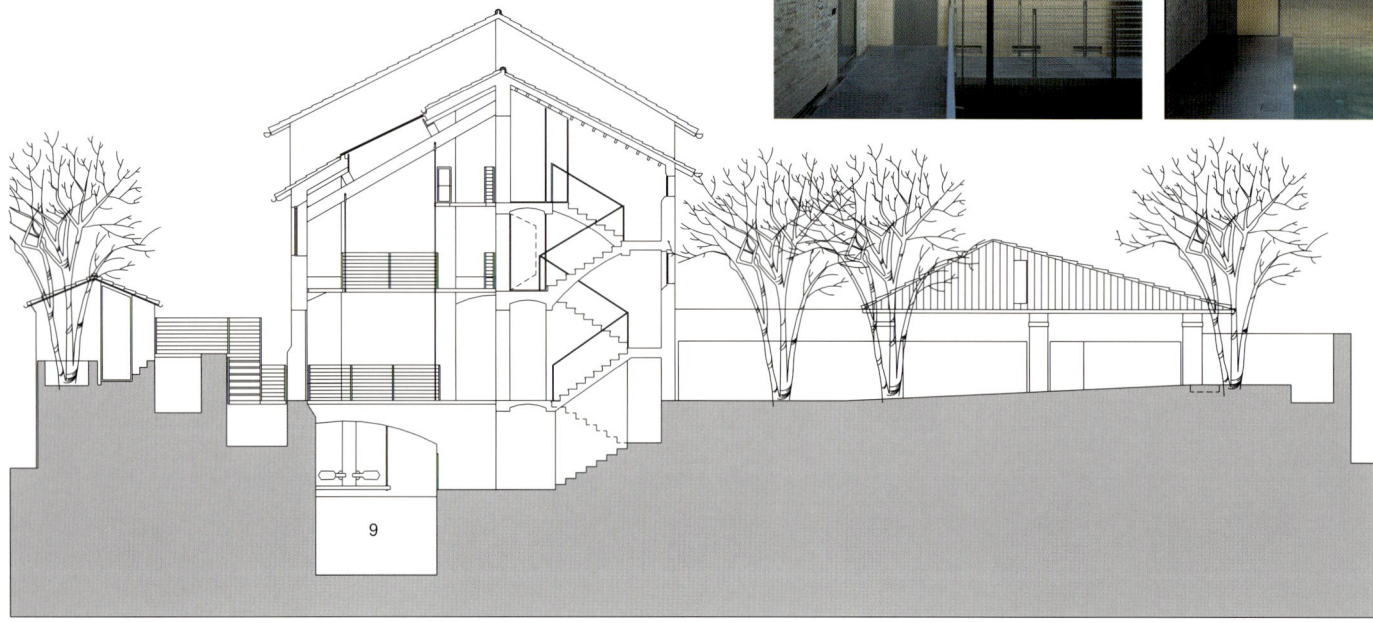

Cross section

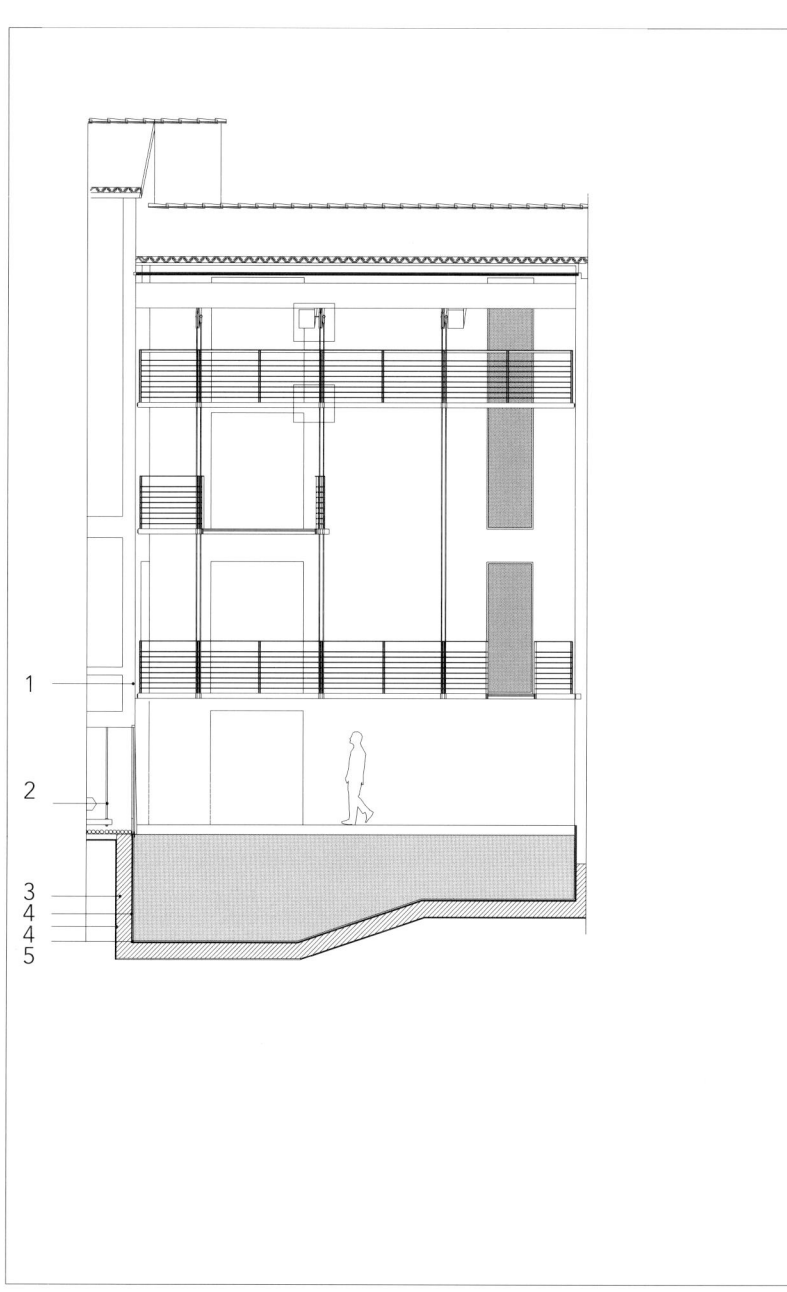

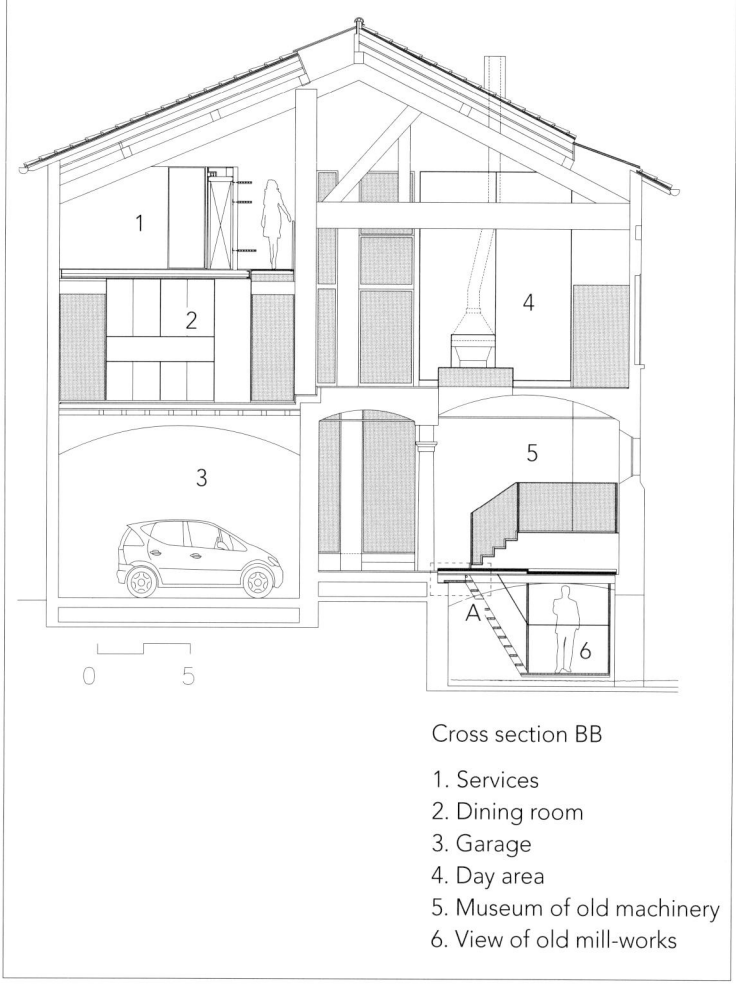

Cross section BB

1. Services
2. Dining room
3. Garage
4. Day area
5. Museum of old machinery
6. View of old mill-works

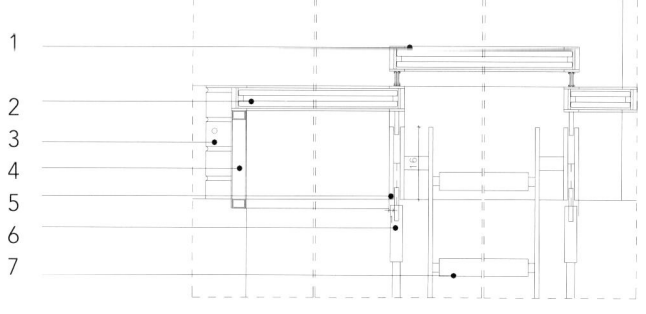

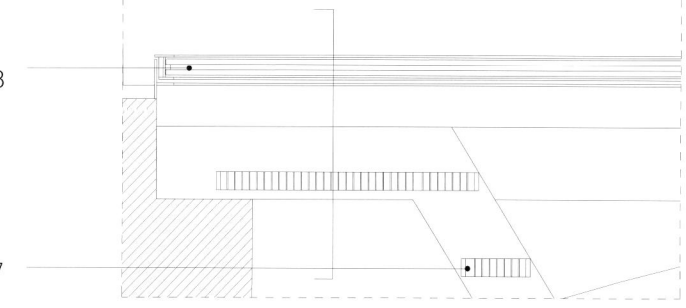

Construction detail A

1. Movable bulkhead for accessing staircase
2. Large fixed window, with drawn out stainless steel fixtures
3. Half-brick cladding
4. Insulated loadbearing structure
5. Loadbearing beams
6. Staircase suspensions
7. Staircase in stainless grating
8. Movable, transparent glass bulkhead with stainless
   steel frame for accessing staircase

Non Kitch Group bvba

Architecture and lifestyle

Koksijde, Belgium

Photographs: *Jan Verlinde*

Transforming an old canning factory in Bruges into an impressive loft dwelling was a great challenge with endless attractive possibilities for the designers. One of the most significant features of this scheme by architect Linda Arschoot and designer William Sweetlove, the creators of the Non Kitch Group, was the remodeled roof. It was formed by a lattice structure with a dog-tooth profile supporting a conventional roof. They decided to replace the north sides of each of the parallel roofs with a glazed surface. These skylights greatly increase the natural lighting in the whole dwelling, as in many museums or art galleries. Due to the considerable height of the building (6 m, 20 ft), this intervention also meant that the interior was almost converted into an outside piazza. A large, full-height room open to the exterior occupies the center of the space and is surrounded by a mezzanine that houses the kitchen, the dining room, the bar and the television room. Under this mezzanine, three steps below the level of the living room are the billiards room, the bedroom, the dressing room, the gym and the bathroom that gives directly onto the small garden. The covered pool is located on one side of this exterior space, with an elegant decoration of vertical mosaic strips. An outdoors area also provides better views and enhances the dimensions of the space.

The conservation of the industrial aspect of the building is shown through the use of metal doors, the heating pipes, the separate kitchen, the galvanized iron staircase and the view of the old factory chimney through the parallel skylights of the roof. In opposition to the asceticism of minimalist interiors, the Non Kitch Group consider themselves to be the heirs of the humor and the colorist aesthetics of the Memphis group. One of the premises of this scheme was to generate an appropriate space for viewing the works of art of the private collection of the owners. The furniture is a forceful presence in this dwelling. Designed by Ettore Sottsass, Philippe Starck, Boris Spiek, Jean Nouvel, Norman Foster and the architects themselves, it seems to be made to measure for this spacious loft.

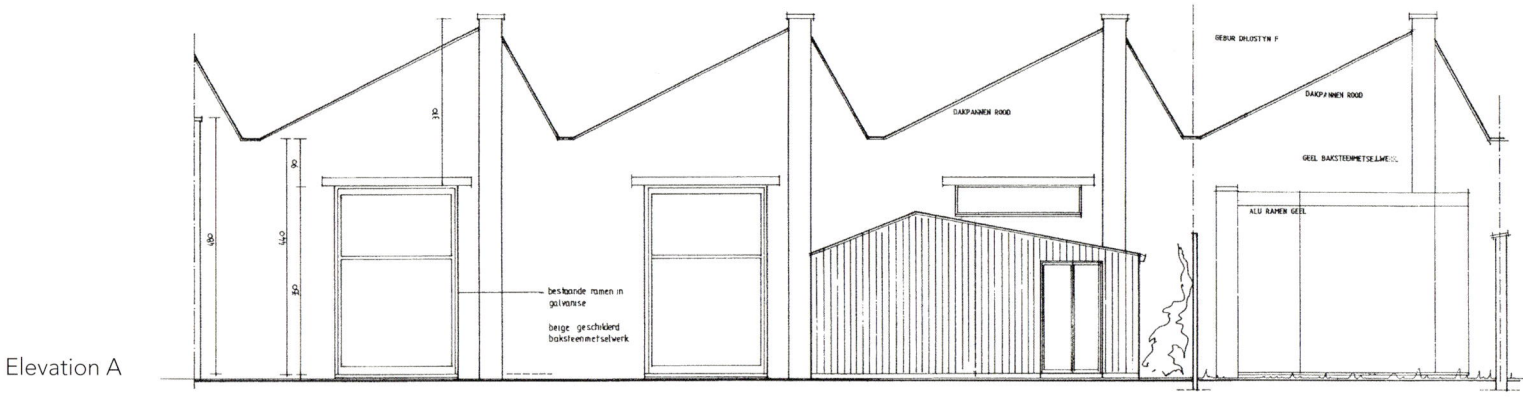

Elevation A

The spectacular glazed ceiling that covers this old canning factory fills the interior with top lighting which enhances the design of the furniture and the objects inside this spacious loft.

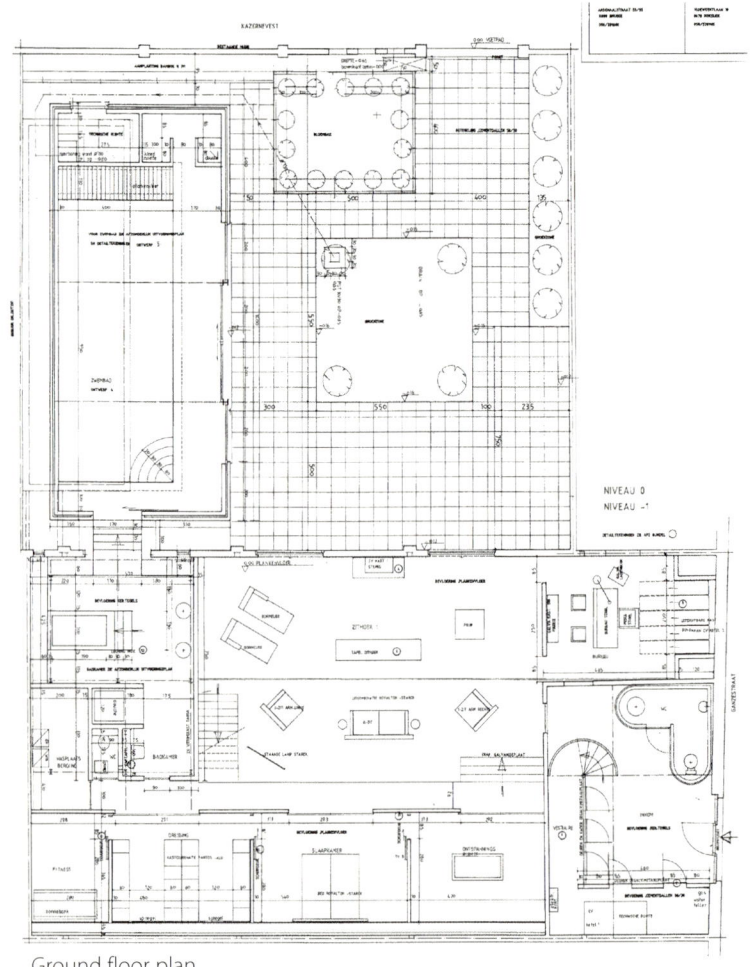

Ground floor plan

One of the premises of this scheme was to generate an appropriate space for viewing the works of art of the private collection of the owners. The furniture is a forceful presence in this dwelling. Designed by Ettore Sottsass, Philippe Starck, Boris Spiek, Jean Nouvel, Norman Foster and the architects themselves, it seems to be made to measure for this spacious loft.

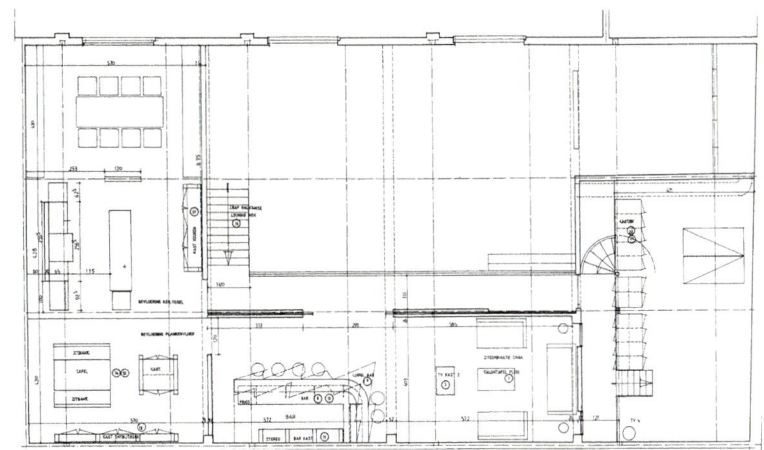

Level +1 +2

The use of bright varied colors to decorate the interior is an amusing and vitalizing feature.

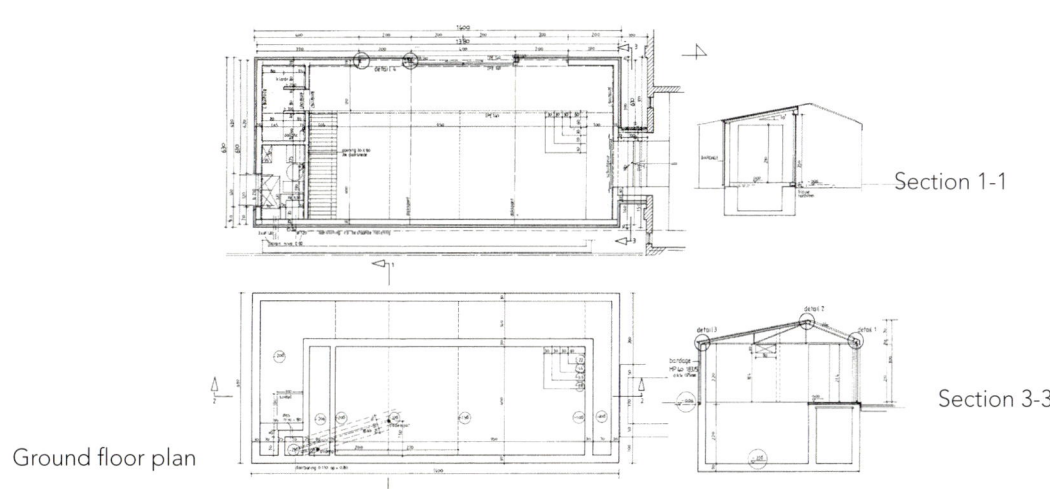

Section 1-1

Ground floor plan

Section 3-3

The metal elements used throughout the loft contribute an industrial air that recalls the former use of the building. The central location of the kitchen and the simple forms are a clear example of the emotions and contrasts sought by the Non Kitch Group.